TREASURE HILL

TREASURE HILL

PORTRAIT OF A SILVER MINING CAMP

W. Turrentine Jackson
with a foreword by Joseph V. Tingley

University of Nevada Press
Reno & Las Vegas

Treasure Hill: Portrait of a Silver Mining Camp was first
published by the University of Arizona Press in 1963.
It is reprinted by arrangement with the author.

University of Nevada Press, Reno, Nevada, 89557 USA
Copyright © 1963 by The Board of Regents of the
Universities and State College of Arizona
Foreword copyright © 2000 by University of Nevada Press
Manufactured in the United States of America

Library of Congress Cataloging-in-Publication Data
Jackson, W. Turrentine (William Turrentine), 1915–
Treasure Hill : portrait of a silver mining camp / W. Turrentine
Jackson ; with a foreword by Joseph V. Tingley.
p. cm.
Originally published: Tucson : University of Arizona Press, 1963.
Includes bibliographical references and index.
ISBN 0-87417-361-2
1. Frontier and pioneer life—Nevada—White Pine County. 2.
Mining camps—Nevada—White Pine County—History. 3. Silver
mines and mining—Nevada—White Pine County—History. 4.
White Pine County (Nev.)—History, Local. 5. White Pine County
(Nev.)—Social life and customs. 6. White Pine County (Nev.)—
Social conditions. I. Title.
F847.W5 J3 2000   979.3'15—dc21
00-008004

The paper used in this book meets the requirements of American
National Standard for Information Sciences—Permanence of
Paper for Printed Library Materials, ANSI Z39.48-1984. Binding
materials were selected for strength and durability.

University of Nevada Paperback Edition, 2000

First Printing
00 01 02 03 04 05 06 07 08 09
5 4 3 2 1

To my mother,
Luther Turrentine Jackson
and
the memory of my father,
Brice Hughes Jackson

CONTENTS

ILLUSTRATIONS

FOREWORD

W. Turrentine Jackson's *Treasure Hill* is more than the story of a solitary mining camp; it is the account of the short life-span of an entire mining district. Jackson's work was the first historical account of the White Pine rush, and it remains today the only study of the frantic activity set into motion by the silver discovery high in Nevada's White Pine Range. There are a few descriptions of the mines and the geology of the district in technical publications, but these by their very nature do not tell the human side of the White Pine story. Jackson makes heavy use of contemporary newspaper accounts to tell his story, and he methodically covers each aspect of mining-camp life, from community water problems to competition between rival stage lines to provide passenger and freight service to the isolated communities of the district. He tells of visiting theatrical troupes, camp-wide epidemics, and the constant search for more silver ore. Jackson documents this search for ore, the sad endpoint of almost every mining camp anywhere, in his account of the Eberhardt and Aurora Mining Company, the London company that entered the district in 1870 after the boom was over and for the next fifteen years persistently invested in a fruitless search for ore.

The boom-bust cycle of mining was played and replayed throughout the American West in the late nineteenth and early twentieth centuries. Discoveries were made and mining camps grew up around them in California, Nevada, Idaho, Montana, Colorado, Utah, Arizona, New Mexico, and even in Alaska and Canada's Yukon. Some of these camps, like Virginia City, Nevada, grew into permanent towns, while others—most others—flashed up for a few years, then passed on to be forgotten. At even the best of the camps, the mines didn't last that long. The best ore found on Virginia City's Comstock Lode was mined out within twenty years of the Lode's discovery in 1859. Mining limped on at Virginia City until the mid-1900s, but the town was really in decline after

1880. Other Nevada camps suffered similar fates. Austin, in the Reese River District, flourished between 1863 and 1887, a period of only twenty-four years. Unionville, in the Humboldt Range, was discovered in 1861, peaked in 1864, and was on its way out by 1880. Pioche and Bristol in Lincoln County, Belmont in Nye County, Tuscarora in Elko County, and Taylor in White Pine County all flourished during the short period between 1860 to 1870 and 1880. Records of the time show that each of these camps produced silver for fifteen to twenty-five years, then faded.

Some mines closed because of the low price of silver. Priced at about $1.33 per ounce since 1850, silver in 1873 began a slide that took it below $1.00 per ounce by 1886 and down to 60 cents per ounce by the turn of the century. This caused economic crises in mining camps throughout the West.

The most commonly recurring theme of decline in the mines, however, was simply the lack of ore. The high-grade, enriched silver ore found near the surface in many of the camps was quickly mined out, leaving only subeconomic roots to be followed by die-hard miners. Rarely could these low-grade remains be mined profitably—at that time, equipment and mining techniques had not been developed for what we now refer to as bulk mining, and open-pit mines were not known.

White Pine, the isolated mining district located on the side of Treasure Hill in a still-remote area of east-central Nevada, perhaps best represents the mining camp of the nineteenth-century West. Even in a time of boom camps with short life-spans, White Pine was an anomaly. It could have been like any other camp in the Mountain West, but it turned out to be one of the richest and shortest-lived of them all. The entire community structure was built in a little over one year, as if the rarified air at the near-9,000-foot elevation of White Pine caused the progression of life in the camp to accelerate. Discovered in 1867, the best was all over by the end of 1869. During those first three seasons, over $6 million in silver was dug from White Pine's mines. The deposits were fabulously rich, most lying within five to sixty feet of the surface. The mineral did not extend to depth, but as in many other western camps, hopeful miners kept searching and digging for years after the bust set in. Nothing like the original bonanza ore was again found, and White Pine quietly died.

When W. Turrentine Jackson's *Treasure Hill* was published in 1963, the camps of the White Pine District had been ghost towns for almost eighty years. Just the previous year, however, in 1962, a gold discovery was made near Carlin in northern Nevada that set off a new prospecting wave across the state. The Carlin deposit was large, and the gold was "invisible"—that is, it could not be seen with the eye or recovered in a gold pan and had been missed by the nineteenth-century prospectors. In the frenzy that followed, every historic gold and silver camp in Nevada was looked over and re-evaluated. White Pine did not escape. A modern geologic study of the district had been published in 1960, and supplemental work in 1970 pointed out that the miners of the 1860s may have looked in the wrong place for more ore. If there were undiscovered silver deposits remaining to be found at White Pine, the deposits would be shallow, hidden under barren rock off to the side of the old mines rather than beneath them where all of the searching was done previously. Also fueling interest, metal prices began to rise. Except for two years during World War I, silver had remained below $1.00 per ounce from the beginning of the century until 1962, when the price rose to $1.09 per ounce. Silver topped $5.00 per ounce by 1978, and during 1980 it averaged slightly over $20.00 per ounce. The price did not stay at that level for long, but it was long enough to bring miners back to White Pine.

By 1977, Treasure City Mines, Inc., had put together over three hundred mining claims in the center of the White Pine District and announced plans to construct a mill to treat eight hundred tons of ore per day from two underground mines and an open pit. Mindful of the water problems that the earlier miners had faced, the new company constructed a 950,000-gallon reservoir at Hamilton to support their planned milling operation. Since only scattered shacks and crumbling foundations remained at the historic camp, an area was cleared above the old Hamilton townsite to park mobile homes—the modern version of a mining camp—for company workers. In 1979, the holdings of Treasure City Mines were acquired by the Gold Creek Corporation of Ely, Nevada. The Gold Creek Corporation produced about 1 million ounces of silver from White Pine during 1979 and 1980, almost entirely from old dumps left from the historic mines. Very little of this silver came from new mining. During the same time, Silver King Mines, Inc., another Ely-based company with holdings in White Pine, hauled a few truckloads of dump material to its silver mill at Taylor, east of Ely. In 1983, more work was planned for White Pine by a Canadian company. This time, a heap-leaching operation was visualized that would recover about 250,000 ounces of silver a year from dumps, tailings, and some new ore. Nothing came of this venture, and another company faded from the White Pine scene.

Today, White Pine is again deserted. The late-nineteenth-century Hamilton townsite is marked by scattered piles of timber from historic mills and a few remnants of brick-and-stone walls with handsome arched doorways from the old town's business district. Signs of the late-twentieth-century operations are a little more obvious. A new metal mill building is still standing, but it is empty of equipment. The mobile-home mining camp was left behind, but vandals have left the trailers unfit for any future use. The reservoir contains water and appears to be usable, if anyone is ever again interested.

However, White Pine's silver mines have the same problem to face today that they did following exhaustion of the bonanza ores in the early 1870s—no more silver ore, and now even the old dumps are gone. It is of course possible that someone, a geologist or a prospector, will arrive in White Pine with a fresh insight and find what has eluded others for the past 130 years. When this happens, there may be a new rush to White Pine.

Joseph V. Tingley
Nevada Bureau of Mines and Geology

PREFACE

Treasure Hill is the story of the rise and fall of one of the significant, but little-known, mining districts in the western United States. On the crest of this hill in eastern Nevada, over nine thousand feet above the sea, the discovery of rich silver ore led to a rush in 1868. The White Pine Mining District was quickly organized, and in time a new Nevada county was created with that name. The boom lasted only two seasons. Capital essential for the continuation of mining and milling operations came from Great Britain. For over twenty years the British doggedly pursued their goal of making White Pine a prosperous mining district. Millions of dollars were poured into the declining community in the 1870's and 1880's. Finally, this underwriting of the economic life of southeastern Nevada drained away the financial resources of the English investors; in a spirit of unbelieving disappointment they withdrew and left behind ghost towns.

Almost twenty-five years ago Russell R. Elliott suggested that the rush to White Pine was "probably the shortest, most intense one in the history of the West." This account of regional history is presented, with extensive detail, as a case study of the social, economic, and political life of a typical mining district in the western United States. In fact, the rush to White Pine, its short season of glory and excitement, followed by sudden decline typifies the pattern of development in the vast majority of mining districts in the West, even in Nevada, much more than the events at the Comstock Lode in Virginia City. These countless, short-lived, booming camps consumed far greater human effort and suffering, more money and equipment, than the occasional successful mining district that gained lasting fame. The relative silence about the failures in the Mining Kingdom has resulted from a lack of records. Men have a tendency to

[1]

forget rather than record disappointment and failure, so the story of the average camp has not won much space in old men's memoirs. The historian must look elsewhere than in personal correspondence and in reminiscences for the evidence.

Indeed the story of Treasure Hill is worth the telling, for it is a chronicle of men, and a few women, who shared The American Dream, a concept combining a belief in the law of progress with the right to pursue success and happiness. They were willing to make the big gamble, wrestle with the inhospitable environment of the wilderness, labor long and hard for success, and finally lose. The miner, with his problems of obtaining housing, food, wood, and water, of guarding himself against the bitter cold of winter, against fire, and robbery by his neighbor, is the hero of our story. Above all else he sought material success, but readily abandoned his work to participate in political activities as one who had a stake in society. Like men everywhere he found release in recreation, and in an isolated mining camp dominated by males, the emphasis was on the physical rather than the cultural. In the end there arises the inevitable question: To what extent was White Pine a part of the frontier experience of the nation? Did Hamilton, the county seat, have problems and triumphs similar or dissimilar to those of the county seat in the agrarian Midwest or on the Great Plains? Did the Nevada miner display personality traits like those of the pioneer dirt farmer and other types on the frontier?

White Pine, in many ways, was a classic situation, not only epitomizing the mining districts throughout the West, but also providing an untouched area where those forces and trends shaping the destiny not only of the West, but also of the nation came into full play. Geographically small and well encapsulated in time, this mining camp provided a microcosm of the turbulence and change in the United States after the Civil War. The life of many a community lay in the balance as rails replaced wagon roads as a part of the Transportation Revolution. So the pattern of migration to White Pine, coinciding in time with the final stages of the completion of the first transcontinental railroad, was determined by the location of terminal towns built by railroad construction crews. The final salvation of the district, after its days of glory, depended upon the construction of a branch railroad that never came. No matter how rich and bountiful the ore, ultimate success depended upon the mechanical facilities to reclaim the precious metal. The technical genius of the nation, expressing itself to the fullest in the Industrial Revolution, found ample outlet in seeking milling and smelting improvements for the mining camps as well as improved machinery for the urban factories of the East. In the 1870's, the British, confident of their technical superiority, often employed mining and industrial engineers at home to design and fabricate machinery, even entire plants, to be transplanted across oceans for use in such places as White Pine.

The plight of the Nevada miner was not unlike that of the railroad and factory worker elsewhere in the nation. Labor unions of the 1870's were self-improvement societies with emphasis upon fraternal, social, and humane functions. To deal in matters of economics, to challenge the views of the capitalist was to court disaster and possibly suicide. Yet the capitalist had earned his right to dominate the decade for he had mobilized the natural resources to build the nation's wealth. To do so he had taken big risks of a type unknown to the twentieth century entrepreneur. When investment capital ran short, as it did in the 1870's, the nation looked abroad for support and the British provided it lavishly. The role played by the British at Treasure Hill, furnishing development capital, managerial and technical skill, was repeated in mining districts in every state that formed a part of the American Mining Frontier. They expended far more in underwriting the mining industry of the United States, and assisting it along the road to maturity, than they took home as profits. In fact, the British invested their pound sterling in every major economic endeavor in the western states. Their American experience provided a major chapter in the story of the migration of British investment capital throughout the world in the nineteenth century.

Two unique collections of historical records in the Bancroft Library of the University of California have formed the basis of the investigation. At the height of the boom, 1869, the White Pine Mining District had three daily newspapers, two of which lasted only a few months. From the broken files of these mining camp newspapers, edited by journalists with extensive experience in mining communities of California, western Nevada, and elsewhere, the historian cannot fail to sense the spirit of the times, boom or bust. The papers of the Eberhardt and Aurora Mining Company, Limited, and its successors, entrusted to William Miles Read, the last general manager of this English enterprise in the United States, are also available in the Bancroft Library. Through its Program for the Collection of Western Americana in Britain, the public records of all the British enterprises in the White Pine District have been procured on microfilm.

As an expression of its interest in state and local history, the American History Research Center at Madison, Wisconsin, sponsored this research and writing project. My continuing study of the British contribution to the mining industry of the American West, of which *Treasure Hill* is a part, has been aided by financial grants for travel, cartographic and clerical assistance by the American Philosophical

Society and the Social Science Research Council. The University of California has consistently supported my research endeavors as a member of its faculty. I am indebted to my friend and fellow-historian of the West, Rodman Wilson Paul, for consultation in the preparation of the final draft of the book.

At the University of Arizona Press, Jack L. Cross and Mrs. Elizabeth Shaw have devoted their time and talent in making certain that the greatest potentialities of this study were realized.

W. Turrentine Jackson

Davis, California
January, 1963

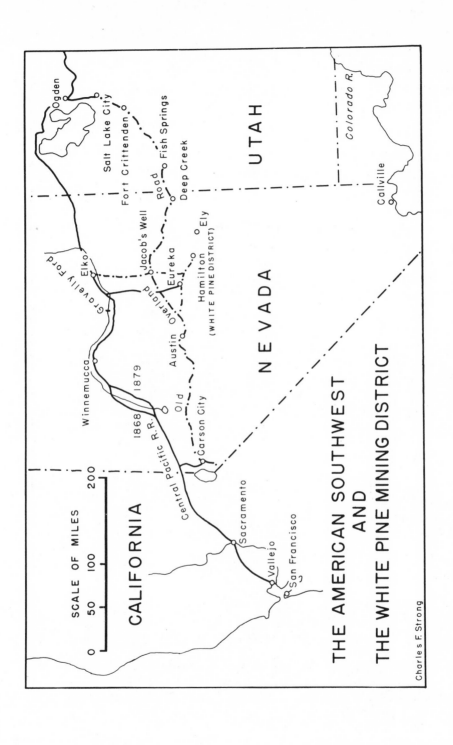

SCALE OF MILES

0   50   100   200

THE AMERICAN SOUTHWEST
AND
THE WHITE PINE MINING DISTRICT

Charles F. Strong

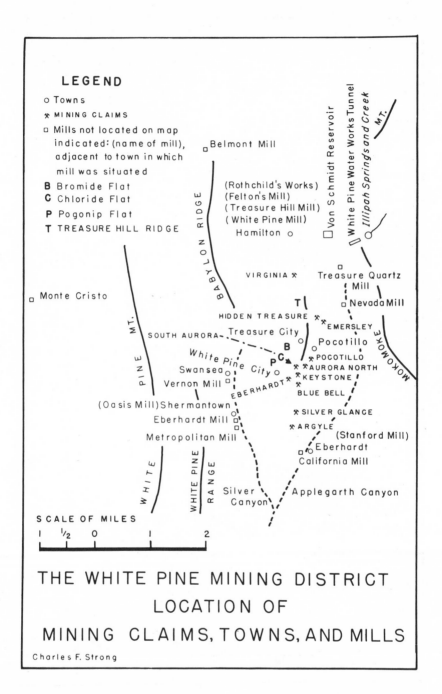

LEGEND

o Towns
✱ MINING CLAIMS
□ Mills not located on map
   indicated: (name of mill),
   adjacent to town in which
   mill was situated

B Bromide Flat
C Chloride Flat
P Pogonip Flat
T TREASURE HILL RIDGE

□ Belmont Mill

(Rothchild's Works)
(Felton's Mill)
(Treasure Hill Mill)
(White Pine Mill)
Hamilton o

□ Von Schmidt Reservoir
White Pine Water Works Tunnel
Illipah Springs and Creek
MT.

BABYLON RIDGE

□ Monte Cristo

VIRGINIA ✱

□ Treasure Quartz
   Mill

T

□ Nevada Mill

HIDDEN TREASURE ✱

✱✱

PINE MT.

SOUTH AURORA

Treasure City o

EMERSLEY

White Pine City

o Pocotillo
B
P C
✱ POCOTILLO
✱ AURORA NORTH

Swansea o
Vernon Mill □

EBERHARDT ✱

✱ KEYSTONE

BLUE BELL

(Oasis Mill) Shermantown o
Eberhardt Mill □
□
Metropolitan Mill

✱ SILVER GLANCE

✱ ARGYLE

(Stanford Mill)

□ o Eberhardt
California Mill

MOKOMOKE MT.

WHITE

WHITE PINE RANGE

Silver
Canyon

Applegarth Canyon

SCALE OF MILES
1   1/2   0   1   2

THE WHITE PINE MINING DISTRICT
LOCATION OF
MINING CLAIMS, TOWNS, AND MILLS

Charles F. Strong

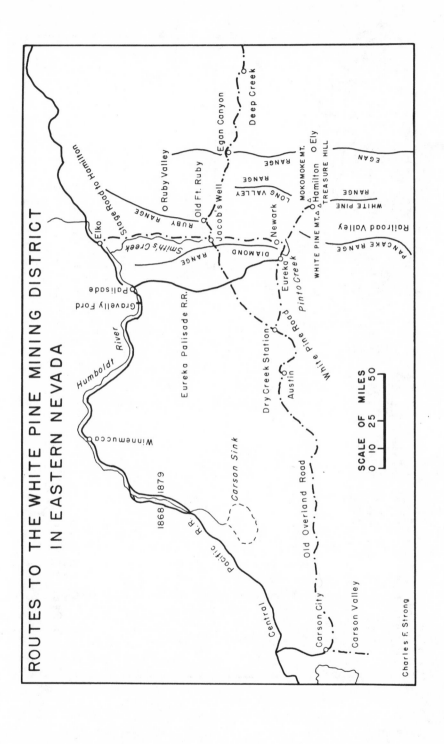

ROUTES TO THE WHITE PINE MINING DISTRICT
IN EASTERN NEVADA

Deep Creek

Egan Canyon

Ruby Valley

Old Ft. Ruby

Jacob's Well

MOKOMOKE MT.
O Ely
WHITE PINE MT. △ Hamilton
△ TREASURE HILL
WHITE PINE RANGE

EGAN RANGE

LONG VALLEY RANGE

Newark

Railroad Valley

PANCAKE RANGE

Stage Road to Hamilton

RUBY RANGE

Elko

Smith's Creek

DIAMOND RANGE

Eureka

Pinto Creek

Gravelly Ford
Palisade

Eureka Palisade R.R.

Humboldt River

Dry Creek Station

White Pine Road

Winnemucca

Austin

Carson Sink

1868 1879

Pacific R.R.

Old Overland Road

SCALE OF MILES
0 10 25 50

Central

Carson City

Carson Valley

Charles F. Strong

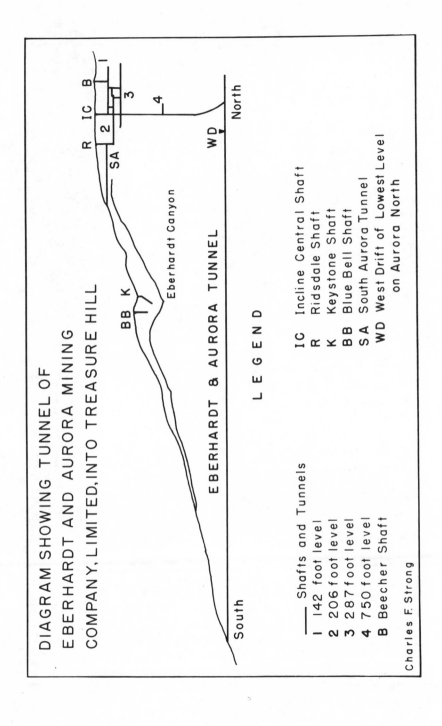

DIAGRAM SHOWING TUNNEL OF
EBERHARDT AND AURORA MINING
COMPANY, LIMITED, INTO TREASURE HILL

South

North

BB K

Eberhardt Canyon

R IC B

SA

2

3

4

WD

EBERHARDT & AURORA TUNNEL

L E G E N D

Shafts and Tunnels
1 142 foot level
2 206 foot level
3 287 foot level
4 750 foot level
B Beecher Shaft

IC   Incline Central Shaft
R   Ridsdale Shaft
K   Keystone Shaft
BB   Blue Bell Shaft
SA   South Aurora Tunnel
WD   West Drift of Lowest Level
      on Aurora North

Charles F. Strong

CHAPTER ONE

THE BEGINNING

One July night in 1867, while asleep in his shanty on the slopes of White Pine Mountain, Albert J. Leathers, a blacksmith turned prospector, was awakened by a noise from his culinary utensils. He observed in the darkness the form of an Indian devouring his dwindling supply of beans. Leathers, whose hard labor and scanty fare had made him neither generous nor peaceful, rose quickly and drove the forager into the night. A few days later a Paiute, known as Jim, appeared, this time to make his peace with Leathers. As a goodwill offering, he presented a piece of silver ore. The blacksmith melted the specimen in his forge and wrought a button of silver through which he punched a hole making a ring that became his prized possession.[1]

Leathers had been with a group of Reese River miners wandering amongst the parched valleys and rugged mountains of eastern Nevada and found evidences of silver on the western slopes of White Pine Mountain as early as the autumn of 1865. Immediately he and his partners called together scattered prospectors to organize a mining district to protect their find. A dozen miners or less, who assembled on the slopes of White Pine Mountain, claimed jurisdiction over twelve square miles, including Treasure Hill. With Robert Morrill presiding, the experienced group agreed to allow each claimant two hundred feet on any "lead" in the district with a right to follow all "dips, spurs, angles, offshoots, outcrops, depths, and variations" wherever they might lead. In accordance with established practice, the discoverer of any new "ledge or lode" was to receive a double claim. After locating his claim, the prospector was required to post a written notice on the ground and record his location within fourteen days. District laws could not be amended, altered, or repealed for two years.

As usual, the recorder was the key man in the mining district, and Thomas J. Murphy, Leathers' partner, was elected to the job for a two-year term. It was his duty to keep a suitable set of books, or at least one book, giving a "full and truthful" record of the proceedings of all public meetings of the miners, to record all claims and notices brought to him in the order of their date, and to make certain new claims did not conflict with the old. The recorder's books were to be open at all times for public inspection, but they were not to be examined except in his, or his deputy's, presence. This would prevent alteration of the records. An unusual decision was made when the recorder was not required to reside in the district. The founders, accustomed to wandering far and wide, recognized that the recorder should be free to leave the district temporarily in search of better prospects during his two-year term without jeopardizing his job, as long as he designated a deputy to act in his absence. If he was unable to perform his duties during his term of office, a successor could be elected at a meeting called upon the petition of fifty miners in the district, provided there were that many around. Notices of the meeting call had to be posted in the district and advertisements placed in the Reese River papers for thirty days so interested parties could arrange to attend. Although equality of opportunity was preserved and the democratic process provided for under specified conditions, the founders had also made certain that they controlled the district, at least until its potential could be determined.[2]

Low-grade ore from White Pine Mountain transported to Austin for reduction produced only from $70 to $150 worth of silver a ton. For this small return, a few miners chose to remain in the district during the winter of 1866. Thomas Murphy, a man of ambition and enterprise, left eastern Nevada in search of capital to build a small mill on the site of the new discovery, thereby eliminating the labor and cost of transporting ore. In San Francisco, he promoted the organizing of the Monte Cristo Mining Company. Unsuccessful in his efforts to raise money in California for this milling operation, he went to Philadelphia with samples of ore from the White Pine District. Here he obtained the funds to hire a trained metallurgist, Edward Marchand, who was to return with him to examine the claims and, if advisable, eventually to build and superintend the Monte Cristo Mill.[3]

Before leaving White Pine, Murphy exercised his right under the laws of the district by naming his partner Leathers the deputy recorder. Leathers was a reliable but unimaginative miner. He guarded the claim and worked hard at extracting low-grade ore. It was while he was awaiting Murphy's return that he had the fateful encounter with the Indian,

Jim. He prevailed upon the Paiute to take him to the spot where the specimen of rich silver ore had been found and their trail led to the top of Treasure Hill. This "hill," six miles in length and three in breadth, was nine thousand feet above the level of the sea. Its easterly and westerly slopes were abrupt and rugged; the northern and southern sides provided a more gentle slope for visitors like Leathers and Jim. The hill was destitute of water, the nearest springs being along the mountain sides of the encircling valley.

As the prospector and the Indian stood on the wind-swept crest they were in desolate and unknown country. The northern and southern boundaries of the state of Nevada were approximately two hundred and fifty miles away, the eastern boundary was slightly less than one hundred miles, and the California-Nevada line three hundred miles distant to the west. Carson City, capital of the state, was due west two hundred and seventy-five miles; Austin, the nearest mining town, was slightly north of west, one hundred and twenty miles away; Jacob's Well, the nearest station on the Overland Road was sixty miles to the north; and two hundred and twenty miles S.S.E. was Callville, head of navigation on the Colorado River.[4] Leathers knew that the White Pine Mountains form one of the larger Great Basin ranges, lying about midway between the Wasatch and Sierra Nevada uplifts. From their vantage point the two men observed that, like most ranges in the Basin, this ridge lay in a north-south direction. Broad desert valleys separated it from the Pancake Range on the west and the Egan Range on the east, where the town of Ely was to be founded. Between the Pancake and White Pine ranges a broad depression was observed full of alkali flats and mud lakes formed by the drainage of the surrounding hills.[5] Prospectors like Leathers, who had roamed the eastern Nevada terrain, had discovered that this broad depression, later known as Railroad Valley, stretched northward a hundred miles to the Humboldt River and southward nearly double that distance. On the westerly margin of the mining district the two men located the bold and craggy outline of a ridge from whence they had come rising from the desert, the crest of which was known as Pogonip, or White Pine Mountain, eleven thousand feet high. Scanning the horizon eastward from this peak they noted the high rolling hills and irregular topography and located the Mokomoke Ridge that formed the eastern and northern borders of their mining district. Treasure Hill, where they stood, was a prominent knoll surrounded by a deep valley with White Pine Mountain to the west and the Mokomoke Ridge to the north and east.[6]

When Leathers returned to his claim at White Pine Mountain after his trip with Jim he decided to call a meeting of the miners in the district to revise their by-laws, in spite of the fact that the two-year time limit had three months to run. The date was set for Saturday morning, July 20. Miners who assembled along the slopes of Mohawk Canyon agreed that it was logical for Leathers, already deputy recorder, to serve as secretary. They brought to the discussion a rich and varied mining experience of the past twenty years. Many had been in California during the rush of 1849, had witnessed the decline of the easily-worked placers within two seasons, and had learned that investment capital or technical experience was essential to develop quartz, hydraulic, or deep tunnel mining. Each miner had soon faced one of two alternatives, either he agreed to work as a hired laborer under supervision for a daily wage or he went on a prospecting trip to find a new claim. Men with the temperament of those who had reached White Pine chose to search for a new El Dorado. Hundreds of California camps with picturesque names, like Git-up-and-Git, Skunk Gulch, Whiskey Bar, and Hell's Delight, had known their days of glory and were later abandoned.[7] Each spring rumors circulated in the mining camps of the Pacific slope that a new discovery had been made, a rush was precipitated, but those who went were disappointed. Then in 1859 two major rushes developed: to the Washoe Mines on the eastern slope of the Sierra Nevada Mountains, and to the heart of the Colorado Rockies. Most of the Californians heading east got only as far as western Nevada. The evolution of the mining industry whereby claims of individuals or partners were consolidated into corporate enterprises capable of investing more heavily in mining and milling machinery came more rapidly in Nevada than it had in California.[8] Some disappointed miners wandered northeastward into the Snake River Valley in search of new placers that could be worked by a man without capital. They drifted from north to south along this stream working every tributary flowing in a westerly direction. In 1860 they made discoveries on the Clearwater Fork, in 1861 on the Salmon, and in 1862 on the Boise. Mass migrations followed one year behind the discoverers on each of these streams. Some crossed the mountains into the Montana region, and starting in 1862 each of the next three mining seasons produced a rush and another boom town.[9]

Other Washoe miners did not have the will or the means to venture as far as the "Inland Empire" on the Northwest but in these same years spread out from Virginia City in all directions. In 1860, one party headed toward the southeast and discovered quartz ledges containing blue streaks indicating sulphurets of silver along the sides of a hill since known as Esmeralda. Five thousand prospectors arrived to establish another boom town at Aurora which was designated by the Legislature of California as the capital of the newly created county of Mono until boundary surveys revealed the town was actually in Nevada.[10] In the spring of 1861 another group of gold and silver mines were discovered in the Humboldt Mountains northeast of Washoe. A half-dozen mining districts were organized during the next two years but the cycle of boom and bust repeated itself within four working seasons.[11] In 1862, ore-bearing quartz was found in the Toiyabe Mountains where the overland mail route crossed the Reese River. The town of Austin was built as a supply center and became the capital of Lander County. The Reese River rush drew off surplus laborers and the more restless prospectors from western Nevada. Although the yield steadily increased from 1863 to 1868, many unsuccessful prospectors once again fanned out from Austin like the spokes in a wheel to make new discoveries and establish a dozen new mining districts: Ione to the southwest, Cortez to the northeast, and most notably Eureka, Ruby, and Diamond to the east where the foundations for Eureka County were laid.[12]

Most of the miners on White Pine Mountain, talking over the revisions in the laws of the district, had begun their mining careers in California, no doubt a few had Idaho and Montana experience, and all had prospected in one or more of the short-lived districts in western or central Nevada. They had participated in the formation of many district by-laws. It was quickly agreed that too many miners who had been in the White Pine area between 1865 and 1867 had located and registered claims, but had done no work to develop them. The miner now had to work his claim to hold it. Within ten days after discovery, claims had to be filed and before forty days had elapsed at least two days' work had to be performed. More important, prior claims were to be valid only until July, 1868. After that date every claim had to be worked each and every year or be forfeited. Only those who stayed and worked were to benefit. Provisions were also made in case a partnership should break up. If one or more of the locators on a joint claim refused to contribute the legal amount of work, a portion of that claim could be segregated for the remaining owners. However, before an individual could designate what part of the claim he wished to hold, the recorder had to give a twenty-day warning to his partners. The term of the recorder was now for a single year. Upon written application of five men, a special meeting of miners could be called, but the majority of the miners in the district had to be present to transact business. At the regular annual meeting in July a majority of those present could settle any problem presented to the group. No person could have the privilege of participating in any meeting unless he was a bona fide miner and had his claim registered and worked as provided in the by-laws. No change was proposed in the basic decision to allow each claimant two hundred feet on the ledge with the right to follow all the meanderings of the vein.[13]

Murphy, upon his return from Philadelphia in September, 1867, heard the exciting news about the recent discovery from his partner, Leathers, and no doubt was shocked to learn that he had concentrated on revising the by-laws of the district, wiping out the rights of those who were not on hand to work their claims, and had failed to locate a new claim at Treasure Hill. Murphy and Marchand joined Leathers and Jim — afterwards known as "Napias Jim" because "napias" was the local Indian name for silver — in a hasty return trip to the hill. On the summit of this bare and windy peak, Jim showed them silver in abundance. Notices of location were immediately made on the ground. For some unexplained reason no work was done to perfect their claim, nor efforts made to record it. Perhaps they acted on the assumption that the working season was over and that winter weather would prohibit further prospecting. Early in November Leathers returned to Treasure Hill, possibly with second thoughts about the advisability of doing a little work, and found that two prospectors had preceded him there. The newcomers had staked out rights to a portion of the same footage selected by Murphy and Leathers and were doing the work essential to perfect their claims under the laws of the district. Rather than quarreling over their respective rights, all agreed to record a claim along the uncovered silver deposit in Murphy's book on November 14 with the last arrivals having precedence over Leathers and his associates. The discovery became known as the Hidden Treasure.[14]

News of the discovery spread rapidly through the Nevada wilderness. T. E. Eberhardt, another miner from Austin, braved the snows of winter to climb Treasure Hill and stake a claim bearing his name on January 3, 1868. Within a few days of Eberhardt's visit a deep snow fell and the inclement weather prevented further exploration.

Experienced prospectors had previously passed over this hill in an unsuccessful search for mines. The geology of the district was mystifying. The float-rock had an unusual appearance. It was generally dark, with a slight reddish or rusty tinge, sometimes yellow, or even black, and looked much like a specimen of limestone colored with oxide of iron. When broken, it was heavy and compact, with a dull, uncompromising lustre, very different from that usually shown by rich quartz.[15] The deposits along the crest of the hill lay flat. Some observers thought that it looked as though a large ledge had been broken off and thrown down on the surface. When it was too late to remedy the situation, all agreed it had been a mistake to allow a discoverer to follow wherever the vein led. The miners, as a group, would have been much better off if locations had been made by the square yard without the right to dig beyond the assigned footage, but to dig as deep as they wished. Fear was immediately expressed that the oldest claimants would swallow up a thousand potential holdings if they had time to work their two hundred feet on the outcropping of ore, and the courts declared them entitled to the pursuit of their find. To check this development, the flats and hillsides were quickly perforated with shafts as soon as the melting snow made possible the return of the miners. There was also an indiscriminate scramble to gouge out the richest ore. From the first, conflicting claims disrupted the peace, and mountain flats became fields of strife and shooting affrays. Lawyers and courts were to be furnished business for many years.[16]

Not knowing the true value of his find, as is often the case, Eberhardt parted with his property during the winter to another group of prospectors — Frank Drake, Edward Applegarth, J. W. Crawford, E. R. Sproul, Lavern Barris, and John Turner — who took several tons of the ore to Austin for reduction. The new owners were amazed and the town was electrified by the news that while some tons produced only $450 worth of silver, others yielded as much as $27,000.[17]

The discovery was timely. After a four-year boom, 1859–63, the Comstock ore chutes near the surface were approaching exhaustion and a business decline had developed. Speculators did not have sufficient confidence in the results of deep exploratory work to pay the calls on their mining company shares. Stocks soon fell to a fraction of their former value; retrenchment and a reduction in miners' wages was the order of the day. In the first two years of the recession, 1864–65, the Bank of California had come to the aid of hard-pressed merchants, mine managers, sawmill owners, and truckers, but when the depression hit in earnest during 1866–67 the bank shed its role as a benevolent patron and began to foreclose. Not only was the Comstock in the doldrums but the great rushes to Idaho and Montana were a thing of the past. In Colorado a new ore deposit and improved smelting techniques were desperately needed to check the flagging economy.[18] The Reese River mines had passed the zenith of production and lesser districts throughout Nevada either had been, or were in the process of being, deserted. In this period of uncertainty and disappointment, the intelligence that exceptionally rich silver deposits had been uncovered on a hill named Treasure somewhere in eastern Nevada was received with wild enthusiasm. The general area was known locally as White Pine. Throughout the mining centers of the American West, old prospectors talked over the advisability of "going to White Pine."

Upon the arrival of the news in San Francisco there was spontaneous rejoicing. The editor of the *Alta California* in his excitement exclaimed, "What a field is here opened for the investigation of the adventurous American people! The gold deposits of California added so many millions to the wealth of the world as to unfix values and affect the whole commerce of Christendom. What California did with her gold, this new Silver Land is apparently about to do with its silver, and no one can predict the extent of the effects which will be produced on our country by the recent discoveries and those to which they will most certainly lead in the coming summer and autumn."[19]

PROMOTING THE BOOM

The element of chance was ever present when new mineral wealth was exposed. At Sutter's Mill in California, at the spring on the slopes of Mount Davidson in western Nevada, at Gold Creek in Montana, and at Treasure Hill in eastern Nevada the discovery was unexpected if not unanticipated. More often than not, new sources of gold and silver were found by a wandering prospector who, due to ignorance and inexperience, had difficulty in protecting his strike. More sophisticated miners managed to appear on the scene shortly, if not instantaneously, buy up the impecunious discoverer's rights for a few hundred dollars, and dominate the district. As we have seen, White Pine was no exception. The men who first discovered silver on Treasure Hill, however, did have the foresight and experience to establish a mining district and systematically register their claims. Moreover, at least two of the partners who first bought the discoverer's rights, Frank Drake and Edward Applegarth, maintained their interests for twenty years until the camp was no more.

All mining rushes are characterized by excitement and over-enthusiasm. White Pine was exceptionally so because prospectors, mill owners, and stage operators throughout the western Mining Kingdom so desperately wanted, and needed, another boom to develop. The typical irresponsible press of the frontier mining towns, printing every rumor full of praise or derogation, was always a significant factor. Where White Pine was concerned, the San Francisco press played a prominent role in building the boom, for publishers and editors cooperated with businessmen in maintaining the primacy of that city on the Pacific Slope. Capital and supplies from the urban center at the Bay had already poured into the Comstock and Reese River districts, and business and financial interests of California were no doubt determined to dominate eastern as well as western and central Nevada if a legitimate mining industry developed. Several rival newspapers and publishing firms released guide books to White Pine and the competition for sales may well have been a factor!

General confusion in registering mining claims, including misstatements of location, clerical errors, lost and misplaced registration books, characterized the first season of most mining rushes. Experience dictated care and caution but everyone was too eager to beat his neighbor to the best location on the stream or to the most valuable section of the quartz lode. Actions and decisions worthy of a week, or at least a day's consideration, had to be made in hours or minutes. Haste meant waste for many at White Pine as elsewhere. As soon as the rush began, there was an immediate need for men, supplies, and equipment. Supply could not meet the demand in the initial stages of the boom, but presently stagers and freighters were legion, and if an organized systematic mining industry did not develop within the first two seasons, there was a problem of oversupply of transportation facilities. Meanwhile, privation and suffering had been experienced by those who could not, or would not, wait.

While snow was still on the ground in the spring of 1868, Nevada miners arrived first and scrambled up the sides of Treasure Hill. The prospector's pick was heard in all directions. Canyons and hillsides were soon adorned with makeshift miners' dwellings, "hanging like bird cages from its rocky sides."[20] All around Hidden Treasure and Eberhardt, new claims were being filed by mid-May. While these early arrivals were excavating almost pure silver from a sizable mass of ore, San Francisco capitalists arrived and, according to a local story indicative of the spirit of the times, offered one group four million dollars for their claim. One of the owners, self-satisfied and cocksure, indignantly replied, "No, when we have taken out enough to pay the national debt, then we will talk about selling."[21]

Immediately following the discovery in 1867, the Nevada State Mineralogist had hastened to Treasure Hill and reported, "Commencing at the point of discovery on the summit, a deposit of ore is traced on the surface due south 1,200 feet, where it disappears under a cropping of limestone for several thousand feet, and reappears on the brow of the southern slope of the hill. It is traced on the surface at this place nearly two thousand feet. It disappears again, but is found further south in several places." The mineral belt was between six and seven miles in length and about one and a half miles in width, and silver appeared to be near the surface of the ground.[22]

While this state official displayed commendable professional restraint in refusing to evaluate the extent and wealth of the silver deposits too quickly, journalists on the scene threw caution to the wind in their reports. Mine owners were also eager to fire the imagination of these newspaper men and exaggerated accounts soon circulated throughout western Nevada and California. Albert S. Evans, editor of the *Alta California,* reported that 2,200 claims had been filed, four-fifths of which were within a circle two miles in diameter.[23] Others insisted that 3,000 claims were located and if only one in twenty proved worth working, White Pine would become the richest mining district in the United States, providing more bullion than the entire previous production of the state of Nevada. Hundreds of miners besieged the nearest assay offices presenting some 1,300 ore samples for testing. The average run was $130 a ton, but many tests produced $5,000, and, as noted, at least one had run as high as $27,000.[24]

The first discovery, the Hidden Treasure, on the north end of the crest of the hill, was quickly stripped during the early spring of 1868 for its entire length of a thousand feet, but only for a width of two to eight feet. At no place was the excavation more than twenty-five feet deep. Picked specimens of horn silver were often worth from $1,000 to $5,000 a ton. Between three and four hundred tons of ore excavated during the working season averaged $160 a ton but reduction costs ran about $65.[25]

Editor Evans reported that the Eberhardt was "beyond any question one of the most rich and extensive deposits of silver ore ever discovered," its name synonymous with that of the cave entered by Aladdin. When the newspaperman arrived at the claim on the southern slope of Treasure Hill, seven hundred feet below the peak, the entrance was closed, but he was given permission to enter by a vertical shaft within a building erected by the owners. "Descending the shaft on a rope we found ourselves among men engaged in breaking down silver by the ton . . . The walls were of silver, the roof over our heads was silver, the very dust which filled our lungs and covered our boots and clothing with a gray coating was fine silver." He estimated that a million dollars' worth of silver was exposed to view, and further observed, "How much may be back of it Heaven only knows. Astounded, bewildered, and confounded, we picked up a handful of the precious metal and returned to the light of day. But for the bars of solid silver since shown us, we would be inclined to doubt the evidence of our senses and look upon the whole scene in the chambers of the Eberhardt mine as the work of a disordered fancy, the baseless fabric of vision."[26] At the end of the year the Eberhardt owners were removing only the ore that would mill at least $1,000 a ton. This was because all available facilities for processing ore were taxed to capacity, and several hundred tons of ore were piled up in storage sheds awaiting reduction. A skeleton force of thirty men was kept at work through the winter months. The fortunate owners had maintained the good will of the camp. The local newspaper editor suggested:

> With such prospects before them, it is evident that the owners of the Eberhardt mine, all being prudent, business-like men, are destined to become millionaires in a very short time. . . . These young men, without being puffed up or betrayed into habits of extravagance by their wealth, have demeaned themselves with modesty, contributing freely towards all needed public improvements, and giving with a free hand to such as have needed their assistance.[27]

In addition to the two mines unearthed in 1867, several additional claims located in the early spring of 1868 proved of real value. On the south end of the crest of Treasure Hill and overlooking the southern slope, not far from the Eberhardt, was the South Aurora Mine. Thirty men were employed to excavate the ore to a depth of twenty to thirty feet. The precious metal was deposited in a series of shallow pockets rather than in a continuous vein. The ore yielded about $185 a ton and the owners were sufficiently pleased that they hastened to build a roof over the entrance so it could be worked throughout the winter months. Just above the South Aurora, nearer the summit, was the Aurora North, that yielded an estimated $440,000 during this first season.[28]

In the neighborhood of the Eberhardt and Aurora claims was the Keystone, where the silver chloride deposit cropped out in incredible richness. According to local storytellers, a prospecting lad named John Turner struck his pick into what appeared to be a mass of dry putty at this claim that proved to be horn silver worth from $15,000 to $25,000. Turner and his father owned one-half interest in the eight hundred-foot claim; Edward Applegarth, the other half. Much of the wealth of this mine consisted of dull yellow-brown colored dust which was run through screens to free it from rocks before placing it in bags for transportation to the mill. This residue was pure chloride of silver. Editor Evans observed that the powder could be placed on an anvil and when struck with a hammer would become metallic silver. He was reminded of "the cheese" produced at the United States mint when silver was being coined — "art having brought the metal to the same point that nature has reached unaided at White Pine." One piece of ore, on the ground, weighed one hundred and forty-three pounds and was appraised at $1,500 in coin. The owners opened a bag of specimens of exceeding richness in the presence of the journalist, and insisted that he take so many samples that he was embarrassed in accepting them.[29] Another journalist who went down into the Keystone shaft placed a value of $8 million on the ore in sight.[30] The owners kept a specimen, about as big as a water bucket and worth $1,500, on display at a local bank.[31]

The Blue Bell claim joined the Eberhardt on the south and its ore was similar. Moreover, no wall or natural line of division separated these two mines from the Keystone. From the time of discovery, the owners talked of the advisability of consolidating to avoid costly litigation over the baffling geological formation. On the spur of the hill about three thousand feet south of the Eberhardt was another valuable claim called the Silver Glance. The ore yielded $110 a ton, but the owners did not work their claim vigorously during 1868. Joining the Silver Glance on the southwest was the Argyle, producing ore worth $200 a ton. In this general area, on the southern slope of Treasure Hill, the West California Company had sunk a "blind shaft" and hit a valuable ledge where there were no surface indications. The hopeful saw evidence here that there was a "well-defined, genuine, metal-bearing fissure ledge" somewhere in the district.[32]

On the eastern slope of Treasure Hill, opposite the Aurora North, was the Pocotillo Mine. In the vicinity of the first discovery at Hidden Treasure were the Emersley and Virginia claims. Ores taken from these mines were unlike those of the Eberhardt and coincident mines. Though all abounded in chloride, these locations were silicious instead of sparry and contained copper and iron in lieu of lead, bismuth, and antimony. The ore here was not so rich, but the owners of the Virginia were offered $80,000 for their claim, or $36,000 for a six-months' lease, and both offers were refused.[33]

Thus the miners who were first on the scene unearthed the most valuable deposits on Treasure Hill. Throughout the mining West, rumors circulated that all these holdings had the potential of the Eberhardt. Men like Tom Cash, ace travelling correspondent of the New York *Herald,* who toured the camp penning vivid accounts for his eastern editors, were largely responsible for leaving this impression. He proclaimed the Eberhardt the richest silver mine in the world with wealth beyond estimate. Already $1 million had been taken out, ore processed daily was worth $6,000, and disinterested parties had insisted he could, with safety, report that $10 million worth of silver ore was now in view at the mine. When his turn came for a visit down the shaft, the New Yorker was permitted to cut out some ore with his knife, so pure it could be rolled between fingers without breaking. He was told the ore he sampled would yield over $20,000 a ton.

> I explained to Mr. Drake, the owner, who politely accompanied me, that if I made assertions concerning the immense wealth of his possessions I might be charged with making false statements, puffing, etc., and to avoid this I would like to take a sample with me of some of the riches [*sic*.] portions. To this he agreed, and, on reaching the ore house, he gave me a piece weighing at least fourteen pounds, much of which would assay to at least $20,000 to the ton, and as a whole, at least $10,000. This piece of rock I have forwarded to New York, so that any one of the doubting Thomases may see for himself.[34]

When Clarence King and the members of his United States Geological Survey visited the camp, the development channel was passing through solid horn silver. His associate, James D. Hague, agreed that the Eberhardt deposit was "probably the most remarkable occurrence of horn silver on record," thus giving official sanction to the flights of fancy by newspaper reporters. Fortunately for his professional reputation he did mention that the depth of the mine was not clearly established.[35] Captain Drake meanwhile bought out two of his partners in the Eberhardt Mine, thereby obtaining a half-interest in this claim. A. P. Stanford of San Francisco, brother of the famous railroad builder, also purchased a claim from another one of the locators for $125,000 and later bought off two conflicting claimants for an additional sum of $32,500.[36]

Throughout most of 1868 all the White Pine ores had to be carted by wagons and pack trains to distant mills. Some went to the Monte Cristo Mill at White Pine Mountain, some to the Manhattan Mill at Austin, and some to the Centenary Company's mill at Newark. Statistics on the ore processed at these mills do not clearly indicate the source. Careful estimates have evaluated the White Pine bullion produced this year at one and a half million dollars. Unfortunately, no confirming tax assessors figures are available for 1868. Shipment records of various express companies and forwarding agents operating out of White Pine indicate that only $500,000 worth of bullion left the district. All agreed, however, that total production was much larger, because some miners entrusted their silver to no one.[37] A large percentage of the bullion also remained in the district to pay wages, buy supplies, and make improvements.

Newspapers and magazines, led by the respected *Alta California*, continued to print many exaggerated reports and idle rumors about the wealth of White Pine. San Franciscans were told that a thousand miners, one-third of the district's population in mid-winter, were extracting 300 tons of ore daily, at an average value of $500 a ton. Thirty-five claims were active. Only a lack of adequate milling facilities held down production. If the deposits proved as extensive as they first appeared, silver production of the district in 1869 was destined to astonish the world. From all available evidence, White Pine seemed certain to be "the wonder of the world for a year or two" and to provide a significant stimulus for business of all kinds on the Pacific slope.[38]

Many thoughtless prospectors, reading such accounts, became victims of "White Pine fever" and rushed to eastern Nevada in the dead of winter only to encounter extreme cold and desolation and to wait four or five months before prospecting and mining were possible. The mining journals urged caution and reminded overly-eager men that the alkaline flats en route would be covered with snow, and atop Treasure Hill the thermometer would sink to 18° below zero. According to reports, three thousand men had already examined the region during the summer and autumn of 1868 and, being practiced in prospecting, they had located all that was "exposed or supposed," either good, bad, or indifferent. These veteran prospectors would renew their work in the early spring knowing what had already been examined, and would explore only new ground and continue to locate claims in advance of the uninitiated and less experienced. Those who insisted upon going were therefore warned not to expect riches from mining, but to content themselves in working for wages or in merchandising. Moreover, they would have to bring their cooking utensils, staple foods to subsist on, and burrow a cave into the hillside to provide a habitation until spring.[39]

Californians in particular were warned not to come until the snow disappeared. Capitalists with limited resources could accomplish little because there was ten times as much money in White Pine, in proportion to population, as there was in San Francisco. Everyone was mad to invest and one typical adventurer remarked, "I expect to suffer some, but I want to get there ahead of the rush." These foolish people were told:

> Bless you, that is just what everybody is saying and has been saying for months. The rush is already going on, and a rush back is bound to come. You can afford to wait in San Francisco, since you can gain nothing by going there now. How many of those who wintered in Virginia City or Gold Hill the first season and suffered everything made a cent out of it? Ask yourselves and reflect. On the 14th of June last fifteen inches of snow fell at Treasure City. It will not be necessary to hurry there with mosquito nets and ice-cream freezers for some time to come. Winter has really but just set in there.[40]

When spring came the rush was on. "Going to White Pine" was a common expression because White Pine was the new El Dorado. The Carson City *Appeal* announced the departure of the governor, United States marshal, and the clerk of the Supreme Court for the new silver district of eastern Nevada. Many merchants, mechanics, and saloon keepers of the Nevada capital had already settled in White Pine, leaving "an immense array of ghostly signs" as reminders of their former presence.[41] The governor, who had invested heavily in mining claims, declared his intention to make the new district his permanent home. Several state senators and assemblymen had followed his lead.[42] The Austin *Reveille* pointed out that all sorts of teams hauling all kinds of wagons filled with every variety of merchandise were passing through that community from morning until night bound for White Pine.[43]

Miners of Idaho territory were stampeding to White Pine to such an extent that a general exodus was feared. The Portland *Oregonian* wrote, "There is a great deal of talk in this city about going to the White Pine mines, and we learn that a similar state of things exists up the [Willamette] valley, especially in the various towns. A good many will doubtless go from Oregon in the course of the season."[44] The tide of emigration from the mining regions of the Pacific Northwest soon merged with another stream from the East. Every day during the first week of March from ten to twenty men headed for Treasure Hill were passing the Ruby Station on the Overland Stage route.[45] The Union Pacific Railroad announced that ten thousand reservations had been made in Chicago for passage to the railheads closest to White Pine. "Now the emigrant has not the tedious and lumbering ox-wagon to follow, no prowling Indians to guard against, no dangerous voyage of many months around the stormy cape to make to reach the land of silver, but the swift running car over the iron road will bear him to his goal in comfortable safety," observed a local resident.[46] Under these conditions it seemed inevitable that the rush of '69 would be as great as the rushes of '49 and '59. The migration was actually large enough that those who had estimated in mid-winter that the population of the White Pine District would be 25,000 by summer revised their estimates upward during April to 40,000.[47]

One observer noted that "it is a pitiable, and yet a merry pilgrimage." Many intelligent young men were seen in the mass of humanity, youngsters who had left behind "a worthy mother who not only hopes for, but is dependent upon him, and whose anguish of solicitude the world probably never even suspects." There were sturdy honest fellows driving oxen teams. Young girls in the company of fancy women and exquisitely booted and jeweled "gentlemen" also passed by.

The capitalist, or agent of a capitalist, expensively dressed, was frequently seen. Scientific men plodded alongside of former convicts who had finished serving their time. The attitude of those encountered on the road by at least one traveler was generous and manly, but crude, and all were full of optimism.[48]

Many mining communities in Nevada and elsewhere in the West were deserted or diminished as a result of the rush to White Pine. Industry in Belmont, Nevada had to be shut down because of a lack of laborers. Men of energy and enterprise were simply unwilling to toil for mere subsistence when fellow workers nearby appeared to be gaining wealth and independence, and soon Austin, Carson City, and Virginia City found business seriously impaired by the emigration.[49] All over eastern Nevada the "prospecting mania" broke out and many prospectors tramped across the wilderness in search of other mines as rich as those on Treasure Hill. Experience had taught them that the claims of real value are taken up in the first season of discovery and they doubted that much would be left for those rushing to the area the second season. But the spirit of adventure had taken hold. By the time summer was over the "unexplored regions" in the area between the Wasatch and Sierra Nevada mountains would be few. With each rumor of a new find, there was a flurry of excitement around Treasure Hill, but no new discovery of significance was made.[50]

While experienced miners were preoccupied with plans to get to White Pine and locate a promising claim, other men were directing their attention to the establishment of stage lines to accommodate the hordes that wanted transportation. Freighters and merchants thought in terms of bringing in supplies and establishing a store. Mill men saw the opportunity of erecting a processing plant closer to the mines.

ROUTES TO THE MINES

Rumors and propaganda about the wealth of the White Pine mines had spread so far and wide that during the dead of winter every conceivable mode of transportation was used to reach eastern Nevada. Several stage companies pioneered new routes across wearisome deserts and through high mountain ranges to this latest El Dorado. For example, a veteran stage proprietor of San Jose and Visalia, California, A. O. Thomas, dispatched a stage for White Pine late in December, 1868, via Walker's Pass, Owen's Lake, across the Inyo Mountains and the northern end of Death Valley into Nevada. This stage route was expected to be the regular channel of transportation from southern California to eastern Nevada, but those who traveled it concluded the difficulties of staging were overwhelming. If this route had to be traveled, it was best to come on horseback, leading a string of pack animals. Parties from southern California were strongly advised, however, to head northward to the Central Pacific Railroad and ride the train to the railhead, presently located at Argenta, shortly to be near Fort Halleck, a hundred miles closer to White Pine, and eventually at the town of Elko.[51]

From San Francisco, passengers could take a boat to Vallejo, thirty miles distant, and, if in a hurry, catch the California Pacific Railroad's train there for Sacramento. More patient travelers used the Sacramento River boat all the way from San Francisco to the state capital. From Sacramento the train carried passengers 460 miles closer to Elko where the stage was taken to cover the last 110 miles, making the distance between San Francisco and the White Pine mines 690 miles.[52] Fare to Sacramento was either $2 by boat or $4 for those who used the train; to Elko by rail was $46; and to White Pine by stage was $40; a maximum cost of $90 for each traveler. Meals along the route averaged $1 each so the prospector's trip was not likely to cost more than $110 provided he did not carry more than an allotted twenty-five pounds of luggage. Most men from the San Francisco Bay area remained in Sacramento overnight, caught the 6 A.M. train headed eastward, and arrived in Elko forty-eight hours after leaving San Francisco.[53]

If one traveled from California with his own horse or wagon, the Sierra Nevada could be crossed by any one of a half-dozen roads leading into Carson Valley. From here the old Overland Road was followed to Austin and on to Dry Creek Station, where the traveler left the thoroughfare and headed southeast on a trace to White Pine by way of Pinto Creek. The three hundred miles from Carson Valley to White Pine required from fifteen to twenty days for wagons, but horsemen could make it in a little more than half that time.

Earliest travelers from the East converged on Salt Lake City and took the Overland Stage via Fort Crittenden, Deep Creek, Egan Canyon, Ruby Valley, to Jacob's Well. Here a transfer had to be made from the overland stage to a local line for the last sixty miles of the journey. As soon as the transcontinental railroad was completed in 1869, many miners and merchants made their way from Salt Lake City northward to Ogden and rode the train west to Elko. Although the stage route between Salt Lake and White Pine was one hundred miles shorter than the railroad route, it was more time-consuming and uncertain.[54]

Parties coming to the White Pine mines from the south experienced little difficulty because the long, open valleys in that direction provided an easy avenue of approach. A new water route was now proposed from San Francisco via the Gulf of California and up the Colorado River, with a road running through the Pahranagat Valley to White Pine. This route not only served Californians but provided an avenue of approach to the people of Arizona and southern Utah.[55]

A Nevada pioneer, who had left the California mines during the hard times of 1867 to "swing an axe" for the Central Pacific Railroad, noted that the building of Elko had just started when news arrived of the silver strike at Treasure Hill. In the winter of 1868 the town consisted of two tents. A veritable "hell on wheels," the railroad riffraff and the roughs en route to White Pine congregated there. Before spring, saloons were legion and the town was a stopping-off place for women of ill-repute on their way to the district. This Nevadan also decided to try his luck in the mines, but ended up working for the San Francisco Sawmill Company that was erecting a sawmill near the Monte Cristo stamp mill on White Pine Mountain.[56] According to the Virginia City *Enterprise* there were ninety tents and three or four board houses in Elko by April, 1869. Forty places sold whiskey, ten were engaged in general merchandise, and in addition there were a drug store, two butcher shops, two harness and blacksmith shops, a dozen lodging tents, many apple and peanut stands, and a hurdy-gurdy house. There seemed to be no work in the village other than the occasional stretching of a tent.[57]

One prospector, knowing that lodging was high in Elko and transportation south almost impossible to obtain, elected to leave the railroad at Battle Mountain and proceed by stage to White Pine. For three weeks he could not obtain a seat on the stage. In despair, he joined a group of Carson Valley miners who banded together and hired out to an immigrant train to drive the wagons and do the cooking. Natural grass was scarce and hay exceedingly expensive when it could be obtained but the group finally reached the mining community with the wagons.[58]

Road travel in Nevada was an ordeal. After one left the Carson Valley headed east or the Salt Lake Valley going west, his route was through barren country, with little wood, grass, or water. There were a number of spots with heavy sand and along some dry stretches of the road water had to be carried. Yet in the spring other sectors of the route were likely to be flooded. The less venturesome who could afford it avoided the overland route and traveled by train. The impression was soon abroad, however, that the roads leading to White

Pine from the railroad towns were in such bad condition that many hesitated to make the journey. Reliable information was difficult to obtain.[59]

There was great suffering among the migrants who hurried to the district in mid-winter and early spring. One prospector reported:

> The weather was intensely cold, the snow deep, and the entire country a bleak, timberless waste, with only here and there a black, rocky peak piercing the surrounding banks and peaks of snow. The thermometer was below zero every day. The stations are exceedingly primitive, nothing but light tents for shelter to man and beast. On the road were crowds of men, struggling along on foot in the direction of the frozen El Dorado. The stages cannot afford transportation for all. At some points we encountered men making their way alone through the wilderness, carrying only a single blanket and a few ounces of provisions.[60]

Travelers were occasionally advised by those who were trying to build a competitive route to the silver land to use these longer and less adequate roads. The Belmont *Champion* warned miners, whether on foot, on horseback, or in wagons, to continue along the main-traveled road between Hot Creek Station and Duck Water Station where water and grass were available, rather than by-passing the latter place along a new and sandy road, said to be five miles shorter.[61]

BOOM TOWNS

Every man attracted to Treasure Hill had to find a place of abode. Earliest arrivals secured shelter in natural caves at the foot of the northern side of the hill 8,000 feet above sea level. In May, 1868, a settlement known as Cave City had been laid out. The next month the first frame structure, a saloon, was built to accommodate the thirty residents, and the town's name changed to Hamilton. By mid-winter this town, consisting primarily of tents, board and cloth shanties, and cabins of rock, brush, and earth, had become the stage and express depot from which supplies were transported by individuals to their mining location. The population had grown to six hundred.

From Hamilton a graded road wound up Treasure Hill to the settlement of Treasure City which was just below the crest, 1,200 feet higher. This toll road was two and a half miles long, but those not hauling loads and preferring to hike up the hill had only a mile and a half to go. Treasure City was exposed to the full sweep of the winds on the summit and was without water except that hauled up from Hamilton and sold at eight cents a gallon. Nevertheless, anywhere from eight hundred to a thousand miners resided there the first season; it was in the heart of the mineral deposits and destined to grow.

A third community, Shermantown, originally known as Silver Springs, was located two miles southwest from Treasure Hill with a road from Hamilton winding down a deep canyon to reach it. Because this community of four or five hundred was sheltered from the winds, it was a more endurable place of residence. Permanence seemed assured when two or three brick buildings were erected while the other settlements could boast only of frame or canvas.[62]

Three-fourths of a mile north of Shermantown, on the road to Hamilton, another milling and smelting town known as Swansea was built during the winter months. A Swansea furnace had been erected here by an experienced Welsh miner who had placed a similar smelting works in successful operation at Oceana in Humboldt County, Nevada. Numerous springs were located in the vicinity, and tunnels were being run for draining them, thereby insuring plenty of water for several anticipated mills. The sanguine looked forward to the time when Shermantown and Swansea would be connected by a continuous street lined with houses and business settlements. Once the village of Swansea began to flourish it became a bone of contention between Treasure City and Shermantown. Both claimed jurisdiction when time came to assess taxes. The legislative act setting the boundaries of Treasure City had included a large portion of the city of Swansea within that municipality, but a subsequent act incorporating Shermantown fixed its boundaries to embrace all of Swansea. The latest law took precedence and Swansea was not destined to have an independent development.[63]

In mid-winter, 1868–69, the entire population of the White Pine District was only 2,500 to 3,000, but was increasing at the rate of fifty a day in spite of the severe cold. Very few women had found their way to the district. Promoters seeking development capital had not yet arrived. Although no one needed to be idle, the "whiskey-bloated bummers" were already on the scene. With the rush of the spring, 1869, the population of the White Pine Mining District soared to an estimated 12,000 and daily accretions now were seventy-five to a hundred rather than fifty.[64]

New towns sprang up in hopes of rivaling earlier settlements. In Applegarth's Canyon at the foot of Treasure Hill on the south side, was the town of Eberhardt, a milling center of one hundred, situated on the flat beside the Stanford Mill. In August two merchants had just finished a fireproof stone building that had been stocked with groceries, provisions, and liquors. In addition, there was a "beautiful supply of good water in the town and both white men and the 'Johns' were driving a lively business in packing it on mules and donkeys to the miners up among the hills." At summer's close, four or five buildings were being constructed in this little town. Atop Treasure Hill, White Pine City had been located just south of Treasure City but this community did not flourish. On the east side of the hill, opposite Treasure City, the new town of Pocotillo had been laid out. One traveler noted, "There is one city called Pocotillo that looked very beautiful on paper; its lots were extensively advertised and then sold at auction; some purchasers were found, and one, when he came to look for his lot, found a mountain top which a kangaroo might climb, but not a human being."[65]

Slowly but steadily, Hamilton emerged as the population center of the district. In addition to being the entrepôt and distributing point, the town was selected as the seat of county government. Between March and May the number of frame houses had quadrupled. Rows of solid buildings lined Main Street where there had been large vacant spaces during the winter. Roads were being graded and plank sidewalks constructed. The teamsters had a tendency to drive their "schooners" to the end of Main street and allow the oxen, mules, or horses to graze nearby. Here a "corral town" grew up with a street lined with "breweries and blacksmith shops, chop houses, and lodging houses, whiskey mills and groceries, saddlery shops and clothing booths." Town boosters claimed that there were over five hundred different places of residence and business in Hamilton by actual count, nearly equal to the combined total of Treasure City and Shermantown, and the population was estimated at 3,500.[66] Leading businessmen of Hamilton began wearing stars made of silver from the Treasure City mines.[67] In this atmosphere of promotion and boom, respectable citizenry was becoming disturbed by the presence of several hundred idlers in the center of town, "bummers, who strike you for 'a piece,' drink whiskey, huddle around fires and swap lies from morning until night . . . as thick as flies in August."[68]

New arrivals from the East were not as impressed with the town as the local merchants and newspapermen. Most of the structures in Hamilton were reported to be of light framework covered with canvas; a total of three hundred was the maximum that some could count. "No one is favorably impressed with the appearance of the town when alighting from the stage," wrote one arrival. "There are no adequate sidewalks, and you land in a sea of mud, which becomes worse as you go to the centre of the street, until at times it is almost impassable. Top boots and overshoes must be worn, and no one need pay any

attention to personal appearance here, for it is the last consideration."
Activity and excitement, however, characterized the place.

> The streets are lined with freight of every description; heavy
> teams are constantly coming in with more; buildings are going
> up everywhere; auction sales are taking place in a dozen locali-
> ties; men are rushing about in every direction, corner loungers
> are in crowds, and all is life, bustle and excitement. You hear
> nothing but claims, real estate and rents; these are the all-ab-
> sorbing topics of conversation, varied only by the saying "Come
> take a drink" — an invitation not often declined . . . There is
> enterprise here and money enough to build it up rapidly, and
> when the season becomes more moderate carpenters and masons
> need not be idle an hour; there will be work enough for all of
> them at high wages.[69]

Both Treasure City and Shermantown also met the challenge of the
situation. At least a third of the district's residents were congregated
near the summit of Treasure Hill during the summer months of 1869.
Lumber had been difficult to procure during the winter, but as soon
as weather permitted, mechanics and laborers were seen throughout
the town preparing foundations for houses, and the streets resounded
with the noise of hammer and saw as the carpenters plied their trade.
Scarcity of water demanded more fireproof buildings, when possible,
and several stone structures of two or more stories were under construction. A log stable was torn down on Main Street to make room
for a stone building to house the Bank of California and the office
of Wells, Fargo & Co.[70] Residents insisted that "the people are at
Hamilton, but the money is at Treasure City." Moreover, the "citizens,
like those of Rome, are seen upon the hills, and are not found, like
the men at Hamilton, lying around loose, or sitting on the dry goods
boxes gaping idly, like the white-headed urchins in the county of
Pike. . ."[71] Shermantown had become the milling center with two mills
going day and night and others expected to start up. A three-story
stone building was erected there to house the Masonic and Odd Fellows lodges.[72]

The growth of these three towns, while others launched with similar
humble beginnings disappeared, was not the result of rivalry and the
spirit of competition but the adaptation of the advantages of each lo-
cality to distinct aspects of the mining industry. Hamilton was the seat
of government and the source of supplies; Treasure City provided the
mineral wealth; Shermantown was the processing center. One optimistic
resident commented, "Here are the three cities of White Pine, all within an area of three miles, young, vigorous, and prospering, containing

an aggregate population of from twelve to fifteen thousand, and each sustaining a live, daily newspaper. Match it, if you can."[73]

The federal census of 1870 did not confirm the population estimates published in local newspapers during 1869. The total population of Nevada was only 42,491, three-fourths of which was male. White Pine County listed 7,189 inhabitants and was second only to Storey County in western Nevada, where Virginia City was the center of mining activity. Within White Pine County, Hamilton had a recorded population of only 3,913, Treasure City, 1,920, and Shermantown, 932. However, the population of the district was notoriously fluid for with inadequate housing available many miners fanned out over the district often living in temporary shelters near their claims. Census takers undoubtedly missed many roving residents. Just as the newspaper estimates were high, the census figures were probably low.[74]

The census also indicates that those residents in the district, born in the United States, came either from California or Nevada or from the states north of the Ohio River. There were practically no southern-born men in White Pine. Of course, many of those who were born in the East had not come directly to Treasure Hill from their homes, but had been in the western mining fields for a decade or more. The most significant aspect of the population pattern in 1870 was the high percentage of foreign-born. Ireland contributed by far the greatest number, followed by England and Wales. There were between four and five hundred men from Canada. The European continent was also well represented, particularly Germany, with Frenchmen, Swedes, Norwegians, Danes and Swiss also present. Two other nationalities, always a part of the mining camps of the Southwest, were here — the Chinese and the Mexicans — with the Orientals over five times as numerous. The number of foreign-born in White Pine County was well over 3,000, being somewhere between forty and fifty per cent of the total recorded population.[75]

WHAT OF THE PROSPECTS?

From the first discovery of silver ore on Treasure Hill, everyone asked the same question, "What of the future?" Even the most sanguine admitted that it was still an open question whether there was a true fissure vein in the district.[76] Many believed that all the metal was merely in horizontal deposits of sedimentary origin. The State Mineralogist of Nevada, after three visits to examine the district, thought it impossible to be absolutely certain. He reported that "there are no well defined walls . . . the ore is much broken and is found in masses or beds, rather than in veins . . . they have no regular dips, and sometimes it is even difficult to determine their course." In his opinion it was improbable that there could be a deposit of ore so uniform in richness running for 1200 feet just below the summit of this isolated mountain and a vein not eventually be uncovered. Such a phenomenon had not occurred elsewhere in Nevada. Even the State Mineralogist was convinced that there was but one Eberhardt on earth; in excitement he stated, "There is probably more immensely rich ore in the Eberhardt Mine than ever discovered in the Pacific States — perhaps in the world." In the long run, however, he thought the "inexhaustible supply" of low-grade ores would provide stability for the district.[77] At the close of 1868, San Francisco's *Mining and Scientific Press* cautiously reported that very pure chlorides had been found at Treasure Hill in irregular deposits under a thin stratum of limestone. Chloride deposits were always surface ores and a richness of $16,000 a ton was "nothing very wonderful" for chloride deposits. The extent to which these deposits had been found was what made the discovery notable in the history of mining. The trade journal thought it not improbable that a vein might be discovered leading to greater depths where the ores would assume the character of sulphurets.[78]

The Austin *Reveille* suggested that the Eberhardt mine was "fishbait" leading "suckers" to bite at many a well-concealed hook.[79] From Virginia City came a report from an experienced Nevada miner:

> It is my opinion, after a careful examination of the country, that there is not enough ore in sight, or even in prospect to keep the present population in bread and whiskey for the next two years, to say nothing of the thousands who are preparing to come next spring. That the ores of this district are of a very high grade cannot of course be denied; but the quantity of ore to sustain a large population is not here.[80]

The first major debate in the press was over the quantity, not the quality, of the silver ore. By early spring 1869, the opinion was gaining ground, at least in the western states, that the number of rich mines at White Pine was limited. Many of the adverse reports came from those annoyed by weather conditions that cut their prospecting season short. Another factor was the jealousy in older fields as both capital and labor were drained away. The Virginia City *Enterprise* printed reports to the effect that two or three hundred men were looking in vain for work, and predicted that unless new discoveries were made crowds of barefoot men would be walking out of the White Pine district by fall.[81] The Nevada *Transcript* concurred.

The truth is, White Pine is already running over with people, and every man who goes there with great expectations of wealth soon begins to ask why he sees so little work done, and parties of ten or a dozen leaving for other points every day. "Why don't the people stay and dig, if there is so much silver?" The only answer a White Piner can give is, "The Eberhardt is very rich," and this is the sum and substance of all there is of White Pine.[82]

The Grass Valley *National* reported that the White Pine excitement had reached its height and was subsiding by the spring of 1869. The Placerville *Democrat* advised its citizens to turn a deaf ear to stories about White Pine, not to exchange a certainty for an uncertainty, but to remain in California.[83]

The editor of the *White Pine News* launched a crusade against the district's detractors, admitting that "the different stories told and representations made by the California papers for and against the mining interests of White Pine are enough to distract the reader, and must leave a very confused impression regarding the facts." Some papers had displayed an amazing, contemptible jealousy. All admitted the superior wisdom of urging caution concerning new mining excitements, but wholesale condemnation without facts was reprehensible. "Each grocery keeper or peanut vendor" in California feared losing a customer and the press was coming to his aid by uttering dire warnings. The editor thought it strange that the San Francisco and Sacramento press should attempt to retard the development of the wilderness and discourage investments which would open up a country and provide customers. Western Nevada newspapers concentrated on drawing comparisons between the production of the Comstock and Treasure Hill. White Pine residents replied that the same fears expressed concerning the depth and durability of the silver bearing lodes of the eastern part of the state might well have been expressed of the Comstock for "much longer after its discovery than White Pine has been known."[84]

Newspapers in many western mining towns seized and commented upon every derogatory bit of information received from White Pine. For example, the following letter from one White Piner to his old friend in Silver City was printed in the Idaho *Tidal Wave*:

To come right down to facts, Tom, this thing is not what it is cracked up to be, mind what I say, and I have taken the pains to find out for myself. It is true, if a man is on the rustle he can make money here. Anything in the shape of a location will sell for cash down; money is plenty here, but my advice to the boys is to stay where they are; there is no use talking, there is more wildcat business done here than I ever saw in all my life. Peter and I are rushing around locating, and Jim and George are working for wages

to keep us in grub, for to tell you the truth, of the $2,000 we
brought here a month ago, there is not a cent in the whole family.[85]

On reading such reports, friends of White Pine insisted that while the
hardy, out-of-doors miners were steadily at work and many of them
reaping great rewards, there were other "prospectors" who refused to
leave the fireplaces and stoves of their cabins and expose themselves to
inclement blasts, but preferred instead to write letters that found their
way to the columns of journals that should have rejected them as trash.[86]

The *Mining and Scientific Press* surveyed the numerous "letters
from White Pine" printed in its exchanges and concluded that they
varied widely in tone. Many had been written by men who, finding
all the mining ground in the district taken up, and seeing no way to
earn expenses, jumped to the conclusion, in their disgust, that the whole
district was a "bilk." The disappointed insisted that there were no mines
but a rich deposit or two controlled by capitalists, and that latecomers
were wandering away on prospecting trips. The truth was hard to ascertain in the spring of 1869. However, enough comment had been circulated concerning the small quantity of ore available that many who
had planned to travel to eastern Nevada, on second thought decided
to stay put. Travel on the Central Pacific had fallen off twenty per cent
by mid-April, and the tent makers of San Francisco, "who are furnishing the housing for the summer's army of occupation in our great interior," reported a marked decrease in their orders.[87]

The nature of ore deposits on Treasure Hill led to as much uncertainty and debate as their size and number. Many capitalists visiting
the mines expressed the opinion that the mineral deposits, being horizontal, could not be depended upon. Most men hesitated to invest
until a true fissure vein, like the Comstock, had been located. The
complaint about "surface deposits," according to one White Pine paper
has been "the burthen of the swan song by the outside press of this
State and the indoor prospectors and do-nothings who have come here
expecting to find the shining half dollars already coined, boxed and
labeled for San Francisco or New York." April 20, 1869, the *White
Pine News* announced that the Eberhardt Company had struck an apparently inexhaustible body of ore, one hundred and ninety feet below
the surface, that assayed almost $2,000 a ton. The editor suggested,
"Let us hear no more about 'surface indications' from any source
whatever." Three days earlier *The Inland Empire* had printed a progress
report on the California mine insisting that it was a true fissure vein
as far as could be determined. "The true fissure veins are with us. Of
this the outside world will be convinced when White Pine gets a little

older. As capital is timid, like a maiden, we give it this advice gratis."[88] Scientists were not so certain. Clarence King and James D. Hague had noted that the Eberhardt mine was described as a ledge, or true fissure vein, and held under the laws that applied to that form of deposit, but "whether it is such, or what its true relations are to a fissure vein, is a matter on which intelligent men hold conflicting opinions."[89] Most capitalists ignored the journalists and waited on the geologists to make up their minds.

The uncertainty was reflected in the mining stock exchange in San Francisco. Rumors circulated in the city that owners of the half-dozen leading chloride mines located in 1868 were sacking their richest ores to send them to the Bay area and to New York with the hope of selling out on the strength of what they had found near the surface. White Pine stocks sold very slowly in San Francisco during April, 1869. A financial authority addressed a letter to the editor of *The Inland Empire:*

> Business here in all kinds of stocks is very dull, especially in White Pine stocks. Capitalists here have a large disgust with the manner your mines are managed as our streets and our markets are glutted with White Pine stocks, which savor very strongly of wild cat. An effort ought to be made in your midst to suppress this wholesale swindling of the public, as a few swindling operations have transpired which have destroyed the confidence of operators in one another, and parties will not purchase now until after the most thorough investigation and examination have been made, and the mine proved good beyond a doubt.

The journalist inquired whether the mine owners of White Pine or the sharpers of San Francisco were to blame for destroying the confidence of the investing public. Such charlatans and their bogus claims had to be exposed.[90]

In contradiction to the accounts of lack of interest in San Francisco, the *Wall Street Journal,* reporting on the California Stock Exchange, announced that "White Pine mining companies are forming at a daily average of one dozen." Chicago and St. Louis were already feeling the excitement, but New York, Boston, and Philadelphia had not yet been affected. The *Journal* thought the White Pine District was "replete with wealth," that it presented a fine opportunity for judicious investment of capital; the New York Mining Board was urged to promote the formation of reliable companies.[91] Was this sharp difference of opinion in San Francisco and New York the result of different promotional methods by stockbrokers in the two cities, or had rumors about meager deposits and uncertain veins failed to reach the East? Without dealing with these matters, the editor of the *White Pine News* decided to concentrate his attention on eastern investors and ignore the Bay area. He requested the New York Mining Board to send out reliable men for a tour of inspection. "Now that the railroad is completed, the facilities for travel between New York and White Pine are as great as could be desired; the cost is but trifling, and the chances for profitable investment here are superior to any ever offered in the United States — not even excepting the oil regions of Pennsylvania in their palmiest days."[92] Gentlemen of wealth and mining experience soon began to arrive from the East, and the newspapers lionized every one. Colonel P. C. Rust, who had been both a proprietor and promoter of several successful mining companies at Virginia City and Gold Hill, came directly from New York with authorization to purchase property for associates if he was impressed with the district. Colonel William Sydney O'Connor put in an appearance on the streets of Hamilton, having come directly from London in the interest of British capitalists.[93] A Chicago correspondent reported an insatiable thirst for information about the location and character of White Pine, and urged the preparation of a pamphlet for sale in the East.[94]

The *Mining and Scientific Press* announced in mid-May that the mystery about White Pine was clearing up. So much had been said about this area being "wonderful beyond description," it had figured so largely as the "poor man's paradise," as a "mount of solid silver," as "different from anything ever known or heard of before," and as "the richest silver mine yet known and dremt of in the world," that the publishers had felt justified in voicing slight disapproval. While no informed person could claim that Treasure Hill did not possess some wonderful mines, hundreds of claims had been filed where only vague streaks of chloride appeared and the owners had adopted a "loafing and selling-out policy." Nothing worse could have happened to the district. At last the professional company incorporators, eager lawyers, and go-betweens with shrewd ideas were disappearing. Meanwhile, enterprising businessmen had adopted a policy of extreme caution insisting on evidence of the existence of at least one well-defined ore channel with sufficient reserves for continuous working.[95]

From the viewpoint of local residents, capital was slow to invest. Some offered the explanation that the money market was tight in San Francisco. In reality, the flow of coin to the interior from San Francisco exceeded the shipments of bullion and dust to that city during April and May, 1869, by $1,324,441. As the expected boom of the spring began to lose momentum, the downhearted thought White Pine was destined for depression and ultimate disaster. The optimistic interpreted the economic trends as leading to stabilization and sound business practices.[96]

With typical civic pride, White Pine newspapers continued to concentrate on two interrelated themes — the glories of the district and desperate need for development capital. One editor insisted that not one capitalist who had come to White Pine and conducted a serious investigation had gone away without becoming interested. The "practical or scientific men" whom they later employed to investigate for them often turned out to be ignorant of the region's geology or were rogues. If only adequate milling facilities could be obtained, White Pine would have ten dividends for every one in any other district of the state, and total bullion shipment for 1870 would reach thirty millions. As the growls against White Pine continued to appear throughout the summer months, the editor of the *News* suggested that the aspersions had become so common they had lost their poignancy; he was giving up all attempts to correct the defamers of the district, but to pass them by with a smile of compassion.[97] A Shermantown journalist displayed great understanding of the situation when he wrote:

> Such is the nature of humanity that all do not appreciate everything alike, and some will growl because every boulder of the mountains is not a mass of pure metal and pronounce the country a "bilk." Many persons are so constituted as to be distrustful and dissatisfied with everything new without reason for investigation into its merits, and they grope their way through life in darkness and doubt from the cradle to the grave.[98]

THE FLOURISHING

With rare exceptions, mineral discoveries were made in isolated areas. An immediate, imperative need was the establishment of communications with the nearest population centers to acquire essential food, building materials, and tools. Traces for men on foot and horseback were located within a matter of weeks after the news of discovery went abroad. These paths soon became roads that the stages and freight wagons could traverse. A few prospectors and businessmen who came with the rush experienced financial success and were caught up in the excitement of the camp, but the typical miner, plagued with loneliness and by memories of those he left behind, was eager for news from faraway places. The uncertainties of communication, especially the mails, did not improve his lot. If the weather forced him indoors, as during the winter months at Treasure Hill, his life became almost unbearable. Camps that prospered for as much as a season or two in the 1870's witnessed the development of a crusade to bring in the telegraph to avoid the uncertainties of surface transportation and maintain contacts with the outside, particularly during the inclement weather. Mill-owners, businessmen, and newspaper proprietors all had a vital interest in speeding communications by wire.

Len Wines pioneered in exploring and opening up a stage line into Hamilton and Treasure City both from the mining center at Austin and from the railhead at Elko. He had already earned a reputation as one of the best stage men in Nevada while working as superintendent on Wells, Fargo & Co.'s overland stage. The route from Elko, lying partly through the Newark and Huntington valleys, had adequate water and forage when first located. Ten way stations along the route were established from nine to seventeen miles apart. Wines' business boomed during December, 1868, and January, 1869, but his monopoly lasted for only a few weeks. Early in February, 1869, the Pacific Union Express Company, a subsidiary of the railroad, was carrying passengers and express between Elko and Hamilton. A week later Wells, Fargo introduced double-daily stages between the railroad terminus and the silver district, placing the line in charge of W. S. Cotrell, one of its most able and experienced men. Migrants welcomed this competition for it meant more lively staging, better time, and cheaper fares. Within a month the traveling time had been reduced from twenty to eighteen hours and the president of the Pacific Union Express was considering the introduction of a pony express from Elko.[1]

The stage and express lines soon were engaged in a more profitable aspect of their business — the export of bullion. The first silver brick left White Pine in February, 1869. This block of precious ore, weighing one hundred pounds and worth over nineteen hundred dollars, was sent from a young miner to his aged father in Dayton, Ohio.[2] It was carried by Wells, Fargo which had an enviable reputation of seeing that precious cargo reached its destination and most of the White Pine mining companies entrusted their bullion shipments to that organization.

The ever-popular Len Wines was well aware of the importance of good public relations even during the weeks that he enjoyed a monopoly. Because of continued complaints of over-crowding, inconvenience, and excessive rates charged by the stages, he released a detailed statement to the public about the costs of his operations. On the route between Elko and White Pine he employed four agents, six drivers, twenty-five stationmen, and used one hundred and sixteen horses. The pay of the four agents was $450 a month, the drivers $75 each, and the stationmen $60. These wages and the board of each man, estimated at $40, aggregated $3,800 monthly. The price of barley was 8½ cents in Elko but increased as one left the railroad until it cost 20 cents in White Pine. Hay was $30 a ton at the end of the rails and $400 at the road terminus. Each horse consumed eighteen pounds of barley and twenty pounds of hay daily, making the monthly bill for feed $17,980. Expenses for wagon repair averaged $1,000 a month, and horseshoeing was $464. The federal government assessed a tax of 2½ per cent on the gross income. There were in addition state and county taxes, interest payments, and losses from accidents that sent the monthly expenditure up to $23,844. To meet this outlay, Wines charged passengers $50 for the trip from Elko to Hamilton, and ten passengers was generally a load; the return rate was only $40 and the stage seldom had more than five passengers. He argued that unless he could obtain income from delivering packages by express, he would be operating at a loss. The mining community appreciated such forthright accounting and favored Wines' stages when possible, but this did not keep them from rejoicing when competition forced a reduction of the rate for incoming passengers to $40. The Treasure Hill newspaper displayed a sympathetic understanding by reminding the community that Wines had been carrying the mail free of charge for several weeks as a public service and suggested that he should receive compensation if he were to be subject to federal taxes.[3] Wines survived the friendly competition with the Pacific Union Express and with Wells, Fargo throughout the winter months, but when it became clear that his former employers were preparing to drive him out of business before the spring rush of 1869, he merged his interests with those of the Pacific Union Express. Once again Treasure City citizens were reminded that he had displayed energy and business sagacity in operating the first stages and was deserving of the patronage.[4]

During March and April, 1869, it appeared that every staging outfit west of the Rocky Mountains had made plans to operate a line into Hamilton. Every mile of railroad completed meant less business for the stages. All were trying to establish feeder lines for railroads and the routes to the White Pine District were exceptionally promising. The Austin *Reveille* announced that a tri-weekly stage line between that town and Hamilton would be established under the management of the Wilson Brothers.[5] Woodruff and Ennor, of Virginia City, dispatched three six-horse Concord coaches, carrying twenty-one passengers, to Hamilton. A large string of horses was taken along to be used in their new passenger service between Elko and White Pine.[6] Hill Beachy, who operated a successful stage line between the Central Pacific Railroad and the Idaho gold fields, departed for Nevada the first week in April with a hundred head of horses to stock a line between the railhead and the silver land. His stages now ran from Silver City, Idaho to Winnemucca; he planned to reroute into Elko, thus forming a continuous line from Boise City, Idaho to Treasure City, Nevada. Idaho's *Tidal Wave* thought that, of all the stage lines on the continent, Hill Beachy's would then be second only to the route between Salt Lake City and Umatilla, Oregon.[7]

The same week that Beachy left Idaho for White Pine two well-known California stage men, A. O. Thomas and Warren Hall, started from San Jose with four coaches and one hundred and thirty horses, with the intention of establishing a line between Elko and White Pine. These men had formerly operated a line between San Francisco and Gilroy, California, but the completion of the railroad along their route ended their enterprise. As we have noted, Thomas had first thought of establishing a stage route down the San Joaquin Valley of California, through Walker's Pass, Owens' Valley, across the Inyo Mountains, and the northern end of Death Valley into Nevada, but one trip in the winter of 1868 had proved the geographic obstacles overwhelming. He and his partner now planned to transfer all their coaches and two hundred and fifty head of horses to eastern Nevada.[8]

Leander Swasey, also of San Francisco, was preparing to put another daily line of coaches on the route from the railroad. He had used his horses to haul coal to the Bay area from Mount Diablo, but the railroad had also destroyed his business in California. Captain Paine, who had been running a saddle train between Elko and Hamilton during the winter months, was now ready to put coaches on the route.[9] Another staging outfit, known as Hughes and Middleton, were periodically running coaches between Elko and Hamilton as early as March, 1869. The next month they instituted daily service. The line had been newly stocked and the owners promised that their stages would always arrive in time to connect with the trains. The editor of the Hamilton newspaper suggested, "Those wearing wigs are advised to take seats on the inside."[10]

Thus within a period of six weeks, seven new stage lines had been put into operation to compete with Wells, Fargo and the Pacific Union Express in conjunction with Len Wines. The Wells, Fargo organization had not been idle during this period of increasing competition. Most of its equipment from the Overland Road had been transferred to the short route between the Central-Union Pacific Railroad termini. As the rail lines were nearing completion the need there was small and temporary, so the idle stock and coaches were to be used in transporting passengers between White Pine district and various railroad points. For the accommodation of eastern travelers, one line was to run from Hamilton through Ruby Valley to intercept the railroad along the Humboldt River. In addition to the direct route to Elko, a third line was to run to Carlin, twenty miles farther west.[11]

The Hamilton-Elko stage road was also shortened and improved during early spring. The so-called Gilson turnpike had been built on higher ground along the north side of Railroad Valley from Newark, lessening the distance fifteen miles as well as by-passing the alkali flats along the old road in the center of the valley. Stations had been built every twelve miles. Gilson kept a corps of five men on his road constantly grading and cutting sagebrush. All of the stage lines transferred their stock and equipment to this new route.[12]

White Pine citizens complained continuously throughout the winter and early spring about the inadequacy of the offices of the major express and staging companies. When the Pacific Union Express doubled the

size of its headquarters in Hamilton at the end of May, the local newspaper commented "it is now nearly large enough for a stage office."

> The offices of this company and Wells, Fargo & Co.'s express are
> doing the largest business of any offices in the State, and they are
> both chucked into little rooms hardly large enough to make a
> comfortable peanut stand. The powers who rule should at once
> procure suitable buildings for their business, and not compel the
> people to climb over the heads of one another to get their express
> matter.[13]

By summer the requested improvements were being made. In Treasure
City the office of Wells, Fargo had been moved to "elegant rooms" in
a new stone building along Main Street jointly constructed with the
Bank of California and reputed to be the most costly and substantial
structure in eastern Nevada. The headquarters was fifty-six by twenty-
two feet with a ceiling over fourteen feet high. A vault had been constructed with double iron doors each having a combination lock. Coins
and valuables were stored in a large iron safe inside the vault so "it
would take an expert burglar about four months to strike chloride
there." Sleeping quarters for employees were available in the basement;
the appointments of the reception room were "of the San Francisco
order."[14]

The spirit of competition expressed itself in attempts to reduce
travel time of the one hundred and twenty-one mile ride from the railroad to Hamilton. During the first week of April the trip was usually
made in twenty-four hours. Len Wines' stages were beating Wells, Fargo
coaches regularly. *The Inland Empire* reported, however, that "Wells,
Fargo men give a wink and say that what happened between Reno
and Virginia will transpire here after a while." The newspaper expressed
a desire to record any races that might occur.[15] From all appearances
Len Wines was destined to beat his former employers all summer, but
all knew that in the end short drives, good stock, and plenty of feed
would decide the contest.[16] The next week the Wells, Fargo stage made
the Elko-Hamilton trip in seventeen hours and twenty-five minutes, the
best time ever made on the road.[17] Competition now became keen and
for a few days the stages of Wells, Fargo, Len Wines, and Hill Beachy
usually arrived in Hamilton neck and neck.[18] The editor of the *White
Pine News* commented on developments.

> Staging is becoming quite interesting, now that several lines are
> established and the coaches' speed and accommodations are all that
> can be desired. Racing has become established. Yesterday we an-
> nounced that the stages of Wells, Fargo & Co. arrived four hours
> in advance of the Pacific Union, or Len Wines' line. For this we

have an explanation that it occurred from the unfortunate breaking of a thoroughbrace which required a few hours to repair. But
yesterday Len Wines' line triumphed coming in an hour and a half
ahead with a splendid Concord coach drawn by six horses and
carrying thirteen passengers.[19]

Competitive racing did not stop at Hamilton, but continued on up
the hill to Treasure City. The two rival express companies kept ponies
awaiting the arrival of the Elko stages at Hamilton. From the outset,
the Wells, Fargo pony usually arrived at the hilltop with its parcels
fifteen minutes ahead of the Pacific Union.[20] The press followed this
development with interest:

> Each day, about Express time, hundreds of apparently interested
> spectators gather along Main Street to witness the arrival of the
> ponies. All seem to take a lively interest in the racing, and no
> matter which pony puts in the first appearance, a hearty round of
> applause is certain to be given by the crowd. Each of the great
> rival companies, of course, have enthusiastic admirers and friends,
> yet it is seldom that we hear of any betting. If the ponies should
> arrive "neck and neck" for a few trips, we have no doubt the
> excitement would extend to the mines, and cause a general sus-
> pension of work for a few hours each day, about "pony time."[21]

By May there were approximately twenty stages and express wagons
carrying passengers between Treasure City and Hamilton. All of them
had a profitable business charging three dollars for the round trip,
two up the hill and one down. On one downward journey, a coach
carrying eight passengers inside and seven on the outside was upset
at a sharp turn in the road near the Mammoth Mine. Travelers on
the outside were pitched down the mountain side about forty feet, cut
and bruised, but not seriously hurt. Everyone took the mishap in good
spirit; the driver was absolved from all blame, and no one criticized
the proprieter for not having securely coupled coach and team.[22]
A reporter for the New York *Herald* traveled the Elko-White Pine
route in April, 1869, on a Wells, Fargo stage with a select party of
three gentlemen from San Francisco and three from New York, intent
on business dealings. Aroused for breakfast at 3 A.M., the party loaded
before sunrise in a mud wagon, "a cross between a stage and an ox
cart, an indescribable sort of vehicle, such as was used on the Overland
route before the railroad did away with all such machines and rendered
traveling tolerable." The travelers soon discovered that they would be
weary and sore long before reaching their destination. Rains had made
the road a mass of mud, heavy teams had cut deep ruts into it, and
cold weather had frozen the ground making it a series of hills and

hollows. Seven miles from Elko they reached the first station for changing horses, but the stock appeared worn, thin, badly fed, and subjected
to exposure in the rudest of stables. The distance between the first and
second stations, fourteen miles, was one of the longest on the route.
Here the journalist resolved to sit outside with the driver rather than be
thrown violently from side to side and have his head jolted against
the ceiling of the stage.

> He was not like those I met on the Reno and Virginia route, but
> a dried up, little chap, that looked only fit for a mud wagon, and
> four inferior horses. He was, however, better than he looked, and
> his descriptions of his trials and vicissitudes while driving on the
> Overland route east of Denver were quite interesting. "I have been
> very lucky myself, sir," he said, "with the Indians; but I have
> passed more than once dead bodies that hadn't been scalped half
> an hour." I thought then an Indian would hardly dare to take his
> scalp, for he appeared such an insignificant specimen of humanity.[23]

At Jacob's Well station, sixty-seven miles from Elko, the stage met the
old Overland road and the wires of the Western Union Telegraph Company. By 8 P.M. they arrived at the door of the Wells, Fargo office in
Hamilton. The journalist was certain that he was "the sorest, most
tired, wornout individual that ever alighted from a one hundred and
thirty mile mud wagon ride." He insisted, "I am almost a cripple, could
scarcely drag one foot after another, and was thoroughly disgusted in
every way, although our trip had only been fifteen and a half hours —
the quickest on record."[24]

Those who traveled by stage did so at their own risk. During the
rainy season of late March, one of Hughes and Middleton's stages,
creeping through a muddy section of the road forty miles out of Elko,
suddenly dropped down and stuck fast. Passengers were summoned to
give a lift. A gentleman from Virginia, John Baldwin, took his position
and was lifting with all his might when the stage started up suddenly,
knocking him off balance and causing him to fall in the mud. The rear
wheel of the coach passed over his body lengthwise. Although he jumped
up announcing he was unhurt, symptoms of paralysis were soon apparent. The traveler arrived in Hamilton, twenty-six hours later, his body
revealing a black and blue strip the width of the wheel. The doctor's
comment was that a weaker man, or one addicted to the too frequent
use of ardent spirits, could not have recovered from such injuries.[25]

A Wells, Fargo stage was upset during the first week of June approximately twenty miles out of Elko. One passenger's arm was dis-
located at the elbow, another suffered a severe leg sprain, and the
others were bruised. Among the party were Congressman Thomas Fitch

and his wife. The stage had been chartered for six people, a common practice among the White Pine aristocrats to secure space which otherwise would have been shared by nine persons. The agent at Elko, however, had crowded an enormous load of freight and baggage on top. Before getting to the place where the accident occurred, an additional six persons, passengers or employees, found places with this luggage. Being so top-heavy, the coach had careened over when it hit a rough spot in the road. No one blamed the driver. In fact his presence of mind in checking the team before the stage hit the ground had minimized the danger. If anyone was guilty, it was the Elko agent, but his excuse was that everyone in White Pine wanted his baggage and freight and everyone wanted a seat.[26] The editor of the *Empire* complained:

> More care should be used in loading stages on the road between Hamilton and Elko, for the reason that the road is very uneven and rough . . . the speed with which the coaches fly over the ground require nothing short of a turnpike road to make it safe traveling when 12 miles per hour are made. At each station the drivers go around and feel the axles to see if they require cold water on them to cool them down. . . . A grain of preventive is worth a bushel of sympathy. Stage accidents are becoming too common.[27]

As the rush to White Pine did not grow to overwhelming proportions, the stage operators suddenly realized there were more stages than the volume of traffic justified and a rate war ensued. On April 15, Wells, Fargo announced that it would lower charges from forty to twenty-five dollars for the stage ride from Hamilton to Elko.

This decision was intended "to make staging lively as long as the opposition lasts."[28] The company proposed in mid-May to drop stage fares to the lowest notch that would pay for the feed of horses. Passengers were to be booked for Elko at ten dollars and from Elko to Hamilton for fifteen dollars. The *White Pine News* reacted in a jocular vein:

> We feel like throwing up our hat; for this will have a tendency to divert the whole of the grand rush from the East White Pineward, and enable the thousands of Pilgrims to investigate for themselves the wonderful mineral wealth of the district. Nobody will pass Elko direct while the fare to White Pine is only fifteen dollars.[29]

Len Wines and Company immediately announced a ten-dollar rate to Elko. Hill Beachy then went down to seven-fifty, and Hughes and Middleton dropped to five dollars. Rumors circulated that enterprising citizens could buy a ticket for as little as two dollars. The press observed, "At these rates 'jerk waters' will have to 'clear the track' or else take passengers through 'free gratis for nothing' and board and whiskey them in the bargain."[30]

The result of the rate war was a general exodus from White Pine, with at least sixty-five men leaving Treasure City and Hamilton by various stages on the first morning the new fares went into effect. Many businessmen who had been in the mining camp during the winter seized the opportunity to return to former residences, settle' their affairs, and return permanently. Some men of means also departed hastily without making a thorough examination of the mining regions. Their departure was lamented by the editor of *The Inland Empire* who suggested the community would be better off if some of the five hundred or more non-producers had gone instead.[31] Within a few days a larger number of panhandlers did leave the district and the journalist commented, "We cannot help noticing the changes on our streets since the late reduction in stage fare. Scores of old bummers have managed to beg enough to pay their fare, and have left for elsewhere, having already worn out their welcome in Hamilton." Many of the men had come from Virginia City and there they returned. "We shall send away several more cargoes before we are done, and we ask Dan De Quille to continue to publish their reports."[32]

The community may have benefited temporarily by reduced stage rates, but the transportation companies were damaged, some irreparably. Many of the smallest outfits sold out, others temporarily suspended operations rather than continue to suffer losses. The Central Pacific Railroad brought pressure to bear on Wells, Fargo & Co. by insisting that the stage line buy up the Pacific Union Express or be denied the railroad's cooperation. To gain favored treatment in securing the passenger and express business radiating out of the towns all along the rail line, Wells, Fargo decided to pay the railroad's price for its subsidiary. Len Wines received notice to quit carrying passengers as an agent for the Pacific Union Express. After withdrawing from the competition for two or three weeks, Wines purchased the entire staging outfit of A. O. Thomas and prepared to run his line again. By June 18, the old rate of twenty-five dollars for the trip from Hamilton to Elko had been restored.[33]

By September, the larger express and stage companies had also given up the trip between Hamilton and Treasure City. A new outfit, Kelley & Co. was plying the mountain road between the two mining communities in colorful green stages that arrived and departed promptly each hour. The public considered them a great convenience, the "boys who manipulated the ribbons" were notoriously courteous and obliging, and on occasions the trip took as little time as thirty-four minutes. All express and mail was entrusted to this company's care.[34] As winter approached, Treasure Hill residents concentrated on road improvements to keep lines of communication and transportation open during the snowy, stormy months. Rice and Company, road contractors, built a toll road eliminating most of the heavy grades up the mountain side, and stages and heavy teams hauling ore down to the mills immediately put the new improvement to test.[35]

At the close of September, Wells, Fargo made a surprise announcement that it was withdrawing its stages from the Elko road, would no longer handle passenger traffic, but would concentrate exclusively on express business. A large portion of the company's stock was sold to Hill Beachy who planned to continue running his stages.[36] Beachy and Len Wines then announced that they had formed a partnership and would concentrate on fine stock, improved Concord coaches, and fast time in running a "railroad stage line" during the winter. Thus, two of the most energetic and experienced stage men in the West merged interests to facilitate their survival during the slack season.[37] Fare to Elko was reduced to twenty dollars.[38]

Chief competition came during the winter months from Woodruff and Ennor who ran daily stages. On December 12, Beachy, Wines and Company reduced fares to twenty dollars from and fifteen dollars to Elko, and Woodruff and Ennor countered by lowering the rate to ten dollars each way.[39] These two partnerships struggled for control of the limited passenger traffic during the winter.[40]

The events associated with the staging operations into the White Pine district clearly illustrate the nature of the unbridled competition that characterized the United States in the 1870's. The businessman of limited means who by his ingenuity got in on the ground floor in any new endeavor was not destined to enjoy his advantage long. Profits brought competition, competition led to conflict and price-cutting, and only those with capital reserves survived. When the struggle was over, the prize was often not worth having and might be abandoned in the moment of victory. Only such men as Len Wines, with fortitude and a sense of timing for surrender or consolidation with the enemy, could hope to survive against corporate competition.

There was no official United States mail on a regular delivery schedule into White Pine during the winter of 1868–69. Various stage operators entrusted the letters arriving at the Overland Stage headquarters in Austin or the railhead at Elko to their drivers headed for Hamilton and Treasure City. In March, Nevada's Senator William M. Stewart notified his constituents that the federal government would let a contract for carrying the mail once a week between Austin and Hamilton. Treasure City officials were incensed that their more distant town was not the terminus of the route and that delivery was not daily, or at least three times a week. The local newspaper commented:

> Those who have had our postal affairs in charge in Washington, are open to the severest condemnation for their neglect of duty. The rapid rise of our city may have exceeded their expectations and they remain in ignorance of the importance of this section, but we hope they are informed by this time, and that daily mail routes direct to the railroad as well as to Austin, will be established, and that Treasure City with its six thousand inhabitants, immense business, and rapid increase will no longer be ignored.[41]

Mail deliveries improved with spring when the various companies making daily trips with passengers and express were also authorized by the Post Office Department to carry mail. Wells, Fargo had the bulk of the business. Sunday was particularly exciting and busy in Treasure City because miners, scattered throughout the hills, converged on the express company offices that day to get the mail. Upon arrival of the ponies, the express offices were jammed for several hours, complaints were deep and loud; the public demanded, and soon obtained, an earlier opening hour on Sunday morning.[42]

There was some difficulty encountered in delivering the letters that were received because of the floating nature of the mining population in the district. In May, 1869, Wells, Fargo had over four thousand unclaimed letters on file.[43] During the winter months, *The Inland Empire* resorted to publishing an alphabetical list of the names of over two thousand individuals who should call at the express offices in Hamilton for letters or packages. In this way, the companies hoped to contact parties prospecting in remote points throughout the mountains of eastern Nevada.[44]

Mail deliveries by stage would no longer be of such vital importance once telegraphic communication was installed. At the beginning of 1869, stories circulated that the Western Union Telegraph Company would not attempt to build a branch line into White Pine before spring; it was thought far more likely that the Atlantic and Pacific Telegraph Company, constructing eastward across the state, would reach Hamilton and Treasure City first. George H. Mumford, general agent of Western Union, then made an official announcement that his company had a contract with the Central Pacific giving it exclusive connection with the railroad, so the rival enterprise could not make a connection at Elko to communicate with California points. Western Union promised immediate construction. Men and materials were already on the ground.[45]

In spite of glowing promises, the end of March came and the wires had not reached the district. The Wells, Fargo agent in Treasure City had been instructed to have his office ready to receive messages by March 15, but spring storms intervened; the first week of the month an ox train had left Elko with the wire but had bogged down in mud and had not been heard from. No one could ascertain just when the telegraph would be completed. Western Union now planned to tap its overland wire in Ruby Valley. Material for the telegraph office in Treasure City soon arrived and was put in operation, but continued interference of the weather halted erection of poles to the crest of the hill. However, the company expected the entire system to be working by April 10.[46] Rumors were still afloat that the line of the Atlantic and Pacific Telegraph Company would reach the district in a few weeks and provide two rapid and comparatively inexpensive means of communication.[47]

When it became apparent that Western Union would fail to meet its second completion deadline, the company announced that a Pony Express would be run between the end of construction and the town of Hamilton for the convenience of the public. After pompous advertising in the local paper, the company was embarrassed when the promised pony failed to arrive. *The Inland Empire,* appearing without its usual telegraphic dispatches brought in by stage, suggested that some accident must have befallen either the horse or its rider. When the pony appeared next day bringing only one San Francisco dispatch, the newspaper announced that "the experiment is a lamentable failure, so far as we are concerned." Nor did the second pony arrive in time to make the newspaper's deadline.[48]

All was forgiven a few days later when the telegraph reached Hamilton. The newspaper explained that an operator who accompanied the construction party when it left the main line at Jacob's Well had only been able, with a small portable battery, to take private messages for White Pine residents. Newspaper releases, sometimes amounting to two thousand continuous words, were more than he could handle with his equipment operating in the open country. In commemoration of the event, the Hamilton editor wrote:

> This is a consummation long looked for and ardently desired by all our citizens. White Pine can, on and after to-night, communicate with the East and the West, by the electric wire, without the transmission of dispatches by stage or pony, to some undefined distance. For this our people have to thank the Western Union Telegraph Company, the pioneers of lightning communication throughout these western wilds. . . . Ye who hug warm fires during the pelting storms of winter in this high altitude can little appreciate the difficulties which the workmen constructing this line have endured.

. . . We hope our citizens will welcome the construction party, and the great event of telegraphic communication with the outer world, in a befitting manner to-day.[49]

A community celebration was held and telegraphic dispatches came through for a single day, but that night a terrific storm "prostrated the wires." Workmen managed to elevate the wires on a sufficient number of poles to get private emergency messages through, but the newspapers again were stranded.[50] Telegraphic communication to the mining camp remained uncertain and sporadic.

Obtaining essential supplies was even more important to the community than receipt of news. During the winter of 1868–69 freight charges were twenty cents a pound. With such transportation rates prevailing, clothing and blankets cost about 50 per cent more than in San Francisco. Food prices were exhorbitant. Early in the spring, merchandise of various kinds poured into the district in wholesale quantities. One observer feared the market would shortly be glutted resulting in a disastrous financial loss to those who had paid winter rates for the delivery of goods.[51] By April, relief from high prices was assured when a train of twenty-seven wagons bringing provisions from Salt Lake City drove in.[52] Merchants looked upon this development with misgivings.

This caravan was the vanguard of freighters who were to flood the district. The thoroughfares of Hamilton were soon lined with pack mules, so crowded that it required "generalship of the vaqueros to segregate them" on Main Street.[53] A large number of prairie schooners were also reported "sailing through Main Street" in Treasure City; several blockades resulting from the locking of wagon wheels provoked "not a little swearing."[54] In Shermantown one team of twelve mules, drawing 21,000 pounds, discharged its cargo for a single mercantile establishment. Machinery for the Metropolitan Mill was simultaneously being unloaded.[55] In addition to those outfits under contract to various merchants and millmen, most stage companies carried freight as well as passengers. At times, two-score wagon trains would be unloading heavy cargoes of freight in the streets of the mining camps. Occasionally there were tragic accidents resulting from the hustle and bustle. Treasure City's newspaper reported that a box of several dozen eggs had fallen from a large wagon when it hit a rut in the chief street of town and the "fruit was smashed to atoms." Commented the editor, "We saw no young chickens in the mud."[56]

The most gruelling task for all freighters was the climb up the hill to Treasure City. The steep grade, however, did not seem to deter the most heavily laden teams from making the trip. While supplies came up by wagon, ore from the mines on the hill was usually carried down by pack animals, and in the early summer some four to five hundred of these animals were being used to supply the mills in Hamilton and Shermantown. Some doubted that ore could be conveyed as cheaply by pack animals as by wagons unless pack animals could be turned loose to shift for themselves at night unlike team animals that had to be tied up and fed. All freighting on the hill by animals or wagons was greatly curtailed most the year because of inclement weather.[57]

By mid-summer, freight rates had been reduced to about one-third the winter level. The Pacific Union company charged seven cents a pound from San Francisco to Hamilton and eight cents to Treasure City and Shermantown. The volume of business of this company continued to grow at these rates, and the public thought the company was deserving of success.[58] Fast freight between Elko and Hamilton cost only one cent a pound during a rate war of early September.[59] By the end of October teamsters were hauling freight for the same distance at two and one-quarter cents a pound, making the trip in about ten days. Fast freight companies made the trip in three days and charged two and one-half cents a pound.[60] Business slowed down considerably the next month and only Woodruff & Ennor were handling freight. As a result, shipments piled up at the railroad depot in Elko and by the end of December special arrangements had to be made to get it hauled to Hamilton. Within a two-day period, 190,000 pounds of freight was dispatched for White Pine and the mining community anticipated seeing the streets thronging again with prairie schooners.[61] Merchants continued to pile up large quantities of merchandise in expectation of a booming business in the spring of 1870. Without it, many were certain to experience bankruptcy.

ENVIRONMENT

The life of the miner was arduous, unpleasant, and often unhappy. He lived in a cave or tent until he and his partners could build a crude cabin of boughs or rough planks. For furnishings he constructed bunks by the side of the wall, used a cracker box for a table, and was deemed fortunate if he had a chair on which to sit. He darned his socks, patched his clothes, and prepared his own meager and monotonous meals. The miner's day was long and hard. He was always exposed to inclement weather and this was an unparalleled burden on the heights of Treasure Hill. Few mining camps had hail storms in the spring, high winds and sandstorms in the summer, deep snow and sub-zero temperatures in winter comparable to those of eastern Nevada. Inadequate housing and diet coupled with exposure to the elements caused as much illness in the White Pine communities as elsewhere. Here, also, was the ever present fear of the outbreak of contagious disease.

During the early part of the first winter season there was an unwillingness on the part of White Pine residents to recognize that the weather was any worse than elsewhere and to resent the continuous adverse comments and warnings in the California newspapers to avoid the place until spring. Early in February the snow on Treasure Hill had not exceeded ten inches, except in drifts, and the coldest day had been only four degrees above zero. Optimists insisted that once comfortable quarters were built, comparable to those in the older mining towns, residents could remain throughout the year.[62] Their enthusiasm was dampened a week later when "the roughest storm of the season" arrived, depositing another thirty inches of snow on the ground. Many frail houses, shanties, and tents collapsed during the heavy gale that accompanied the snow.[63] Before the end of the month the temperature had dropped to twenty-three degrees below zero. The oldest inhabitant insisted it was the coldest period he had ever experienced in eastern Nevada, and conditions now seemed to justify the extravagant reports circulated in California about the severity of White Pine weather. Many a miner, ensconced in his cabin, had the canvas roof of his dwelling collapse under the weight of the snow. Some were forced to dig out and bunk with a friendly neighbor until spring thaws.[64] Numerous incidents of suffering were related, but the miners could still laugh at their predicament:

> A singular case of frost bite occurred on the Divide between Treasure and Hamilton during the extreme cold of Sunday night. A jolly fellow, muchly elevated — we are all elevated about nine thousand feet, but he was spiritually elevated — making the journey on that night between the two places, after answering an imperative call, carelessly exposed himself, and upon reaching Hamilton discovered that he was frozen — well, where women could not be. Suddenly sobered by the appalling calamity, he frantically rushed for professional aid, and the medicine man at once applied an abundant poultice of snow. Partial relief was thus given, but it is expected that early amputation will be necessary in order to enable a respectable — chignon. We have heard of weather causing such disasters to brass monkeys, but not to the genus homo.[65]

When the winter continued into April, the miners' dispositions began to show the strain. One miner descended from his claim to visit a friend living in an excessively warm house behind the Pacific Union Express office in Treasure City. Soon after taking his seat he drew his six-shooter and fired a few shots through the side of the building. A crowd quickly gathered thinking another unfortunate man had committed suicide. The visitor stared at the curious, put away his pistol and calmly remarked that he wasn't going to be smothered to death "in such a place as this 'ere."[66]

Hail squalls, accompanied by thunder and sudden, violent wind, became the order of the day in the springtime. Canvas houses blew down in Hamilton and Treasure City, and most of the tents in the district left their moorings.[67] The gales reminded many residents of the "zephyrs" they had experienced in Virginia City and Gold Hill "when wagons loaded with quartz were blown over and stages were swept off a grade sixty feet wide." The editor of Hamilton's *Inland Empire* suggested, "Tents and flimsy houses will do very well in calm and pleasant weather, but when a White Pine zephyr comes along they are of little use — almost worse than none. What a summer may be in this region is hard to conjecture."[68] Ground softened by melting snow and rains at this season occasionally loosed boulders that clattered down the hillside destroying all in their path. One twenty-ton rock was sighted as it began such a descent in time to warn the men working on the La Patria ledge. They fled in haste but all their equipment and machinery was destroyed. The boulder finally struck a cabin, knocking it down and smashing all the contents to fragments, before coming to a resting place. It was reported that a miner with a cup of coffee in hand calmly walked out the door as the rock tore through the cabin.[69]

A severe snow storm developed as late as the end of May. Capitalists from San Francisco, seeking investment, could not endure the climate and departed en masse. The situation was a repetition of events in Virginia City during May, 1863. In disappointment, the newspapers suggested that men who leave pleasant and luxurious homes at sea level should not expect to enjoy the same climate at an elevation of six to eight thousand feet. Nor should they expect the conveniences, in the way of lodging and boarding, that could be found in the cities of the Pacific coast. Many miners feared that capitalists would not return until Hamilton and Treasure City had been rebuilt of stone and brick, and first-class hotel accommodations provided.[70]

Summer was short in the mountains of eastern Nevada. By the third week in August the nights were cold enough for frost. Already "the signs admonish us that the season of 'pogonip' is approaching," wrote a local editor, "instead of thunder showers we may soon expect to see snow flying occasionally."[71] On August 28, 1869, a tornado came out of the west, passed over the city of Hamilton in the late afternoon, and disappeared up the valley eastward. This "destructive hurricane" levelled a dozen buildings; one of the largest mills in the vicinity was a total wreck, eight men were injured, two perhaps fatally.[72] After the middle of September heavy frost whitened the surface of Treasure Hill, covering everything like a sheet of snow each morning.[73] Cold wind began to blow with a vengeance and it was observed that "all a man had to do to get his breath was to stop it — no draw necessary."[74] The temperature in Hamilton dropped to twenty-seven degrees on October 1, and the public school had to be dismissed because no stove had yet been installed in the building.[75] The first snow of the winter of 1869 fell in November, leaving a two-inch blanket of white. Shoshone Jim announced that there would be no more "white fly" for another moon, and Jim was said "by the early Pogonippers to be no slouch for a prophet about such matters." In spite of this forecast, a severe storm struck the next week and another four inches of snow fell.[76]

Everyone recognized that the weather was a hackneyed subject for discussion, but there was nothing as important to the people of the White Pine district during the winter of 1869–70. Since winter weather would mean a stoppage of work in the mines, scores of men would be out of a job. Most mills were supplied with sufficient ore to continue running through the winter, but they would not be able to absorb the unemployed mine workers.[77]

A coating of ice was everywhere on December 21, 1869. The thermometer dropped to one degree below zero in Hamilton and the next day went to six below. Stage drivers reported that, for the first time, the entire road to Elko was covered with snow. Main Street in Hamilton was described as a "perfect glare of ice from one end to the other." Icy streets were hard on teams; footing was precarious ascending the grade, and on the downgrade, locks had to be used, jerking the animals mercilessly.[78] Early in February the wind continued at gale velocity most of the time. On one occasion the Hamilton Opera House was badly damaged when its canvas top was torn to shreds; sets were jerked from the stage and twisted into a thousand shapes. In Treasure City there was complaint about the signs swinging dangerously in the wind above the heads of passers-by on Main Street, and store proprietors were urged to secure them with more substantial wire lest somebody be knocked down.[79]

The great storm of the year broke on February 24. The *White Pine News* reported:

> Yesterday was a wild day on Treasure Hill, and a number of jurymen coming down on foot wandered from the trail and came near to freezing before reaching this city. In the forenoon the storm was furious, and passenger sleighs had to suspend travel

for a time, the roads being covered by drifting snow and the storm so blinding that neither horses nor drivers could see.[80]

The streets of Hamilton were deserted and hot stoves were sought by every resident. Hail and sleet, mixed with snow, fell for forty-eight hours. Many recalled that during this same week the previous year temperatures had ranged from zero to twenty-four degrees below.[81] Several smelting works shut down because snow had made the roads impassable even for pack trains. Many mines also had been compelled to suspend operations because they had no covering.[82] When the snow stopped falling, the thermometer dipped to fourteen degrees below zero in Hamilton and to twenty below in Treasure City.[83] The newspaper editor in Hamilton asked a miner friend from Treasure Hill how deep the snow was; the miner replied, "Well, I don't know exactly, but according to my measurements, it is from two to ten feet according to the lay of the land."[84]

The annual weather pattern in the White Pine District for the years 1868–70 was by no means unusual. Each subsequent year brought with it a bitterly cold winter with sub-zero temperature much of the time, a wet and cold spring that lasted into April and sometimes May, and then a short, hot summer with high winds and dust. For example, in April, 1872, White Piners experienced a snow storm that was described as the severest of the year; twelve inches of snow fell and in places banks were three and four feet deep. Parties starting from Hamilton to Treasure City were obliged to turn back, whether on foot or horseback. Residents in Treasure City were compelled to stay indoors or travel on snowshoes. Storm followed storm and by the month's end the local newspaper suggested "the chances are that people who wish to take part in next 4th of July procession will have to supply themselves with snowshoes."[85]

"The Great Storm" of 1875–76 commenced on Christmas Eve and with the exception of one or two days it continued to snow and blow until the end of January.

> Anyone who has never experienced one of our big snowstorms, accompanied with wind, can have but a faint idea of their severity. To persons who have to encounter them they are not pleasant; the weather is usually very cold when these storms occur, and the cold and drifting snow cuts the face and blinds the eyes. How stage drivers and teamsters, who are compelled to be out in them, stand these storms is a query to us. . . . If a dozen big teams were to go over the roads during these storms, in two hours afterwards all signs of former tracks would be entirely filled by drifting snow.[86]

As a rule, Treasure Hill and Hamilton had more snowy and windy weather than any other part of eastern Nevada, but all the countryside experienced this cold wave.[87]

HEALTH

The cruel and piercing winds of winter on Treasure Hill brought colds and pneumonia, disabling many of the inadequately housed. A few died. Those unaffected insisted it was only the men with weak lungs who succumbed to pneumonia and that a little care would have prevented it.[88] Midwesterners enjoyed the high elevation where there were no fevers and ague. Even so, some with weak lungs contracted tuberculosis and were forced out of the mines. Worst of all, in the coldest weeks of the winter of 1868, when the thermometer was hovering below the zero mark, an outbreak of smallpox occurred on Treasure Hill. Fortunately the afflicted were isolated soon enough to prevent a major epidemic.[89]

Many exaggerated stories of sickness and death in the mining region circulated in western Nevada and California. The Treasure City newspaper set the record straight in January, 1869, by reporting that there had been only five cases of smallpox, most of which had been in Shermantown. Two had been fatal. In the same camp there had been one death from pneumonia.[90]

In the spring, smallpox became a greater danger to the community. A yellow flag nailed to the front door of a house on Main Street in Treasure City was the first indication that the disease had reappeared. The miner who had been stricken wanted to avoid going to the pesthouse, so a single attendant was obtained to aid him in his cabin, closed to all other visitors. Although this patient died, no new cases of smallpox were reported in the settlement on the crest of the hill.[91] Meanwhile, a correspondent from Treasure City had written the Reno *Crescent,* "Plenty of smallpox and pneumonia here; the latter is, I believe, the most fatal. Hundreds are still pouring in here; what in h—l will become of them I am a little dull to see." Hamilton's *Inland Empire,* representing a town temporarily free from disease and illness, replied, "Smallpox and pneumonia are mere bagatelles at this altitude. Silver indications are the thing! If our friend of the *Crescent* had ever opened a Bible he would have made the discovery that the 28th chapter of the Book of Job holds out hope for all who come to White Pine."[92] Two weeks later there were nine cases of smallpox in Hamilton.[93] A lodging-house keeper on Main Street had died; his wife who had attended him during his illness had also become very sick and was unconscious at the time of his burial. Friends reported her illness to be rheumatism, but public opinion insisted she had smallpox.[94] While an important land-title case was being tried before the District Court in Hamilton a few weeks later, one juryman suddenly broke out with smallpox and the other eleven took to their heels. The trial was indefinitely postponed.[95] Fear of the disease was so great that miners on Treasure Hill learned they could protect their cabins from robbers and intruders by raising a yellow flag each time they left home.[96] Pneumonia continued to take a greater toll of human life than smallpox. Some weeks in the spring as many as four deaths were reported. Miners, eager to begin the season's operations, exposed themselves in the cold and snowy weather and succumbed to colds, influenza, and the dread pneumonia. Many young men in their thirties also suffered heart seizures in the high altitude after drinking too much.[97]

Summer heat brought the pesky flies, particularly in Shermantown. When the editor of the *Telegram* complained, a neighborly rival remarked, "We feel sorry for him, and would relieve him if we could. He has to take his lemonade and lager with a genuine fly in it, very much against his will. His ideas are confused, when writing items, by the pesky flies who persist in roosting on the end of his nose. The flies like toddy, and they show their knowledge of men by roosting near where they are likely to find it most frequently."[98] The good humor seemed inappropriate a few days later when typhoid fever broke out in the camp.[99]

White Pine miners attempted, from the first, to care for the sick and dying and thereby protect the community. In the winter of 1868–69 an isolation house was established in Hamilton under the supervision of a newly-arrived doctor, where those "lying low from malignant smallpox" could be kept.[100] When the outbreak of the spring came, twelve business establishments subscribed one hundred dollars each to maintain this "pest house" in Hamilton where there were eleven residents, five of whom were from Treasure City.[101] Noting this generosity, the editor of *The Inland Empire* suggested:

> There is no place in the State where a hospital is more needed than in White Pine. . . . There are hundreds of persons among us who have no one to look after them if sick. Everyone is carried along by the whirl of the excitement, and never stop to think of the sufferings of their fellow men. Let there be a comfortable place provided, and if we have no one to occupy it, there will be no harm done, and should we be visited by any contagion or epidemic, we will have a suitable place for the sick.[102]

The county commissioners soon established a hospital supported by taxation, but the facilities were so inadequate the press proclaimed it "a shame and a disgrace." Following a surprise visit, Hamilton's journalist reported:

> A board shanty in a desert ravine, and a dozen sick or wounded men in the same room foot to foot or side by side, feeling nothing but suffering and sorrow, and seeing nothing but anguish. . . . There is not a sheet or pillow slip in the whole institution; there is not an outhouse of any description adjacent to the hospital; there is no awning, not even a canvas one, on any side of the structure to cool the building or furnish shade in the open air for convalescents.[103]

The commissioners were called upon to remedy the situation. "As it is now, it is a wicked burlesque on what a hospital in a civilized community should be, unless the design of such institutions is to reconcile suffering mortals to welcome death."[104]

NEWSPAPERS

During the 1850's and 1860's mining camp newspapers in California were legion, and numerous frontier editors, with their presses and type, followed the prospectors across the Sierra Nevada and established more newspapers in the mining communities of Nevada than either population statistics or economic conditions warranted. In Virginia City men like Mark Twain and Dan DeQuille gained fame as distinguished mining camp journalists. At Treasure Hill the newspaper men arrived with the first wave of prospectors, and three typical publications were established in Hamilton, Treasure City, and Shermantown. Myron Angel and William J. Forbes were second only to Twain and DeQuille. No citizen was more important to the mining camp than the newspaper editor. The paper he published became the eyes and ears of the community recording daily events for posterity. His files became the primary source of information on the social scene, on economic conditions and political activity. His publication also established contact with the outside world advertising the glories of the isolated camp in metropolitan journals and, in return, learned of national and international events of interest to the hard-working miner. Some editors had the rare gift of making their newspapers the center of intellectual comment or the repository of local literary effort. The historian is forced to conclude that more often than not the newspaper editor supported his community to a far greater extent than the local populace underwrote his enterprise.

"In colonizing a new and remote district, among the earliest wants that manifest themselves among the American people is a local newspaper," observed W. H. Pitchford and Robert W. Simpson, the two proprietors of the *White Pine News* in their first weekly edition at Treasure City on December 26, 1868. The publishers, printers, press and type for this pioneer newspaper in the district, all came from the office of the *Reese River Reveille* of Austin. Before that their equipment had been used to print the *Silver Bend Reporter*.[105] The publishers' aims were twofold; to furnish the public with a useful and reliable journal and at the same time to establish a legitimate and paying business. While national events were to receive attention, the *News* intended to make local affairs its special province.[106] This weekly newspaper was sent to subscribers for $10 a year or $1 a month.

In January, 1869, plans were revealed for the establishment of a second newspaper in the district with headquarters in the town of Hamilton. James J. Ayers, formerly of the San Francisco *Call* and Los Angeles *Express,* and C. A. Putnam, of Virginia City's *Territorial Enterprise,* had purchased the *Daily Safeguard* of the latter community. This elaborate printing outfit was transported overland to White Pine to start *The Inland Empire*. Reports stated that the owners and editors would publish a weekly until spring and probably a daily thereafter. The *News* commented, "This is only about the sixth paper destined for White Pine, and should each one succeed in reaching here it would make the newspaper business *lively.* . . . We would like to leave the people of California one or two papers, and other parts of this State at least one, for we are not avaricious in this respect."[107]

The new venture in Hamilton was delayed for two months and, in preparation for the competition, the proprietors of the *News* announced that Myron Angel, distinguished and witty journalist from the *Reveille,* would serve as editor. In February the paper was increased to a tri-weekly and in March to a daily. Among the residents at Treasure Hill was William J. Forbes, formerly of the *Humboldt Register* and the Virginia City *Trespass,* who had abandoned his profession to join the rush to White Pine and establish a saloon. In May he resolved to return to journalism, leased the *News,* succeeded Myron Angel as editor, and in July purchased Simpson's half-interest in the publication. Under Forbes' control the newspaper was an active advocate of the program of the Republican Party.[108]

Without due appreciation for the magnitude of the task of transporting a large printing establishment to White Pine, the owners of *The Inland Empire* issued a circular letter early in February to the regional newspapers suggesting that exchanges begin on March 1. Winter storms and impassable wagon roads delayed delivery of the press so that the first issue did not appear until March 27, 1869. During the interval the "contemporaries" remitted their papers ignoring the silence from *The*

Inland Empire. Once underway, this Hamilton daily sold for sixteen dollars a year, ten dollars for six months, or six dollars for three months. The owners announced that the paper would be independent in politics and published a statement of objectives:

> With a desire to make the world familiar with the most astonishing mineral development in history, and also with the view of actively assisting in the opening to useful purposes our other vast resources, THE INLAND EMPIRE will be conducted on no narrow or selfish principles. The county of White Pine will find in it a good friend, because it will be a truthful and candid one. We shall advocate the maintenance of the laws, the enforcement of order and the observance of public morality. We shall strenuously insist upon the honest and faithful performance of their duties by the officers of the State as well as those of our own county. We shall fearlessly denounce crime wherever we may find it—in short, THE INLAND EMPIRE will be a JOURNAL OF THE PEOPLE.[109]

On receiving the first number of the new journal, Virginia City's *Territorial Enterprise* commented that this paper, half its own size, had started off with a respectable display of advertising patronage and had every appearance of a prosperous future. As to the claim of political independence, "It will be as independent as two Republican editors of ability will be able to make it."[110]

When the *News* added the word "Daily" to the masthead, prices were raised to match those of the *Empire*. Austin's *Reveille,* under the heading "People Cry for It," announced that its citizens were unhappy because the *News* was so difficult to obtain. The editor responded from Treasure City:

> Lord bless you, dear Reveille, we work with all our might to get off some thousands of copies on our slow hand press, and then for a day and a night in making up our mail, and after all this work and care don't scold us, if, with our present postal arrangement, ninetenths of those we send away fail of reaching their destination. We know people cry for the News, as a child cries for candy, a local editor for his tod, or a belle for a new bonnet, and we do our best to supply all. We have before this spoken of our mail communications, but it is of little use to complain, as in a few days more, with roads completed and routes let, then some regularity and certainty will be obtained. The fault that all exchanges and subscribers are not supplied is not ours, for we faithfully mail it to you.[111]

The Inland Empire announced that a pony express would run between the telegraph station at Jacob's Well and Hamilton to furnish the latest intelligence from all quarters of the world, thereby making this publication "a paper of news." The editors commented, "Such expedition might not be much for *The New York Herald,* but it ought to be considered pretty good for White Pine, which has been heretofore well down in the dark about outside news."[112] The size of the paper continued to cramp the editors, but on April 13, the promise of enlargement was fulfilled. At last winter was almost over, roads were constantly passable, and a continuous supply of paper could be obtained. The owners stockpiled enough to keep the newspaper in business for many months. The publication was now the size of the Sacramento *Union,* generally recognized on the Pacific slope as the standard of a first-class journal.[113] The Reno *Crescent* observed, "This paper [*The Inland Empire*] came to us yesterday greatly enlarged and improved. The name does not look half so pretentious on the enlarged paper as it did in the first copy. It is now a first-class paper, if its editors do commence God with a little g."[114]

By May, the *Empire* was doing well and the owners pronounced it an "undoubted success." On the score of circulation, there was only one journal in the state to which the editors were willing to "lift our castor," and none between the Reese River Range and the Mississippi River. Under these circumstances businessmen could see the obvious advantages of patronizing the Hamilton paper.[115] The newspaper blossomed out with an elaborate new masthead. Months earlier the electroplates had been entrusted by the manufacturer to one of the express companies for delivery, but for some reason they had been misplaced and had meandered around the Comstock Lode for weeks. The paper commented, "The vignette, especially, has been so long on the road, and so late in coming, that we decline to explain why a mine enters one side of the globe and a railroad emerges from the other."[116]

The self-approval of the owners of the *Empire* was, in part, justified according to a reporter of the New York *Herald* who pronounced the *Empire* an excellent daily newspaper published from an office housed in a tent, thirty by sixty feet in size. He informed his own editor:

> In one corner I found one of the editors hard at work on copy for the next day. In another corner, of which a room had been made, the other editor was in bed sick; half a dozen compositors were at work; there was a power press, Washington hand press, two Gordons and rotary, and the proprietors told me that they were run down with job work. Really, I saw nothing more American in character than this newspaper office; and if energy deserves success truly the proprietors of the establishment should make a fortune.[117]

These newspapers were recognized as an integral part of the White Pine community. The public was friendly to the proprietors, editorial writers, and staff. Businessmen constantly sought their good will and support. The editor of the *White Pine News* prepared a local item called "Far-Fetched, but Good" in which he said,

> A man walked into our office yesterday, and deposited several bottles on our table, accompanied by a short note from Perry & Irwin, Main Street, Hamilton. It was a good distance to bring bottles, but we partook without hesitation of their contents — having known Jack and Bene long enough to be satisfied they would not, at this late day, begin to deal in anything but the very best of liquors.[118]

This practice of saloon keepers became routine, and it never failed to produce favorable publicity for the contributor. Commented *The Inland Empire* editors on one occasion, "Biggs made another attempt to have our local muddled last night, by sending a bottle of something to the office, but he failed — because the Bourbon was so put that it only brightened the intellects of our typos. Biggs keeps on Main street, immediately back of our office, and sells none but the best liquors."[119]

The rivalry between the *White Pine News* and *The Inland Empire,* both morning newspapers, induced editor Forbes of the *News to* sponsor an evening paper that would circulate in Hamilton and thereby reduce the advertising and circulation of the *Empire.* He prevailed on Pat Holland, a former carrier and agent who had become dissatisfied with his business relations with the Hamilton paper, to assume the role of proprietor. In June this third newspaper appeared on the scene, the *White Pine Evening Telegram* published in Shermantown. The first page of the initial issue advertised sixteen saloons, seven restaurants and lodging houses, and printed twenty-two professional cards. The paper was actually printed in the office of the *News* on Treasure Hill. By wintertime it had folded. For a month or two early in 1869 Shermantown had a small independent newspaper known as the *Reporter,* published by E. F. McElwain and U. E. Allen with William H. Clipperton as editor. The press was a small one that had been previously used to print the *Nye County News* and the *Mountain Champion* of Belmont, and the paper never gained the influence or reputation of the Hamilton and Treasure Hill publications.[120]

In this contest for primacy in the district, a new verbal slugging match between the *News* and the *Empire* was precipitated on July 10 when *The Inland Empire* ran an advertisement addressed "to business men" stating that it had the largest circulation of any newspaper in eastern Nevada and offered the best and most profitable medium for advertisers. The editor of the *News* countered by stating that the publishers of *The Inland Empire* had no way of knowing the circulation figures of the *News* and were therefore "wilful falsifiers" and violators of the code of respectable journalism.[121] From this time forward the two newspapers agreed to disagree. Greatest acrimony developed over the truth or falsehood of reports of new discoveries, over the amount of bullion produced, about the success and failure of various mining companies, and the decision as to whether White Pine should look to Chicago or San Francisco for capital.[122] The editor of the *News* commented on one occasion:

> The "local" of the *Unreliable,* at the foot of the hill, can not understand why the fact that our bullion statement for the week is made up on Friday and his, the day following, should show a discrepancy in the figures as published by each. As our giddy young friend is not a mathematician we will bring the thing down to his capacity, if possible, by an illustration. . . . The NEWS begins the week on Monday, the *Unreliable* on Tuesday. The Jew claims Saturday as the seventh or Sabbath day, while Christians claim Sunday as the Sabbath. There is a discrepancy somewhere in this business. Do you think you see it Patrick?[123]

At one stage in the fight, the Hamilton paper answered that all but one of the city fathers in Treasure City had left town despairing of its future development, and the *News* immediately reported the presence of two of the "departed" on the streets.[124] The *Empire,* continually poking fun, dubbed the *News,* the *Reprint* because of its reliance, to an extent, upon the more complete telegraphic news coverage at the base of the hill.[125]

LAWLESSNESS

Lawlessness and turbulence characterized the mining camp. Idlers and roaring reprobates hastened to any new boomtown in hopes of making easy money. The law-abiding majority in the typical mining camp sooner or later was forced to call a mass meeting to settle difficulties with the troublesome elements of the community. If this failed, a vigilance committee was organized to restore order. The White Pine District had its quota of gamblers, prostitutes, thieves and confidence men. Residents there held mass meetings on occasions, but vigilante action never became necessary. Nor were the primitive devices of public whipping, exile, and hanging resorted to. Homicides occurred much too often to suit public opinion, but the culprits always managed to escape before the marshal or the community meted out punishment.

Although life was as uncertain and violent in White Pine as in most mining towns, apparently crime was not quite as rampant nor justice as rough and ready. Henry Eno observed the differences when he arrived in "this land of sin, sagebrush, and Silver" after spending twenty years in the California mining camps. He thought there was less "reckless extravagance" among the people of the towns and at the mines than in early day California. There were not as many homicides in relationship to population but more highway robberies were committed.[126]

"Society, properly speaking, there is none, and the community is in a wild, seething, chaotic transition state," observed a journalist visiting White Pine in the first winter following the discovery of silver. The majority of the residents at this time were educated, enterprising, and orderly citizens. But rougher times were ahead. So many undesirables of every type and nationality began to swarm into the camp that they threatened to dominate it.[127]

Idleness among able-bodied men was a continuous and potentially-dangerous problem. The curious and adventurous, many of whom had loafed around Virginia City for years, came to Hamilton to participate in the excitement. Sleeping on the sidewalks soon became a common practice. Men were also seen snoozing comfortably on a bank of snow, while barrels and lumber piles were sought out by the enterprising. Eno stated, "We have here . . . numbers too lazy to work but not too lazy to steal and some too proud to work and not afraid to steal."[128] The city police, who had to look after the health of such men, advised those who found themselves a little top-heavy to strike for home or some other place indoors.[129] Ne'er-do-wells who drank to excess were usually held accountable. The Hamilton press reported the first offense of "John Doe and Richard Roe" who had been found dead drunk on the streets, lodged in the county jail overnight, and then after a severe reprimand, told to "go and sin no more." However, the marshal, who acted as their intercessor, sentenced the culprits to stack a wagonload of wood in back of the jail to pay for their night's lodgings. The local editor suggested, "This is a good idea; if the Marshal had some town property to work these fellows on, it would cure them of such tricks. Our streets would be a good place."[130]

On occasions, intoxicated or crazed fellows went on a rampage and tried to shoot up the town. A favorite pastime of some was to fire three or four shots into a saloon where miners were quietly seated discussing their prospects.[131] In the spring of 1869, fast horseback riding became a real community problem. One drunken man on horseback dashed through Main Street in Treasure City at a desperate pace, without regard for his life or that of the hundreds of other men passing along the crowded thoroughfare. When his horse felled one man and slackened his pace, the rider slapped the animal with his hat and rode on at break-neck speed, until another and then another was knocked senseless. Several men attempted to shoot the madman, but before they could draw he was out of range. A dozen residents lay strewn along the street or prostrate on the mountain grade leading from town before he disappeared over the hill.[132]

Though drunkenness was often the cause of serious accident and injury, the press preferred to report the humorous incidents.

> Yesterday afternoon a couple of inebriated courtesans, named Carrie Anderson and Molly Sheppard, while preambulating about town at an uncertain pace, were simultaneously thrown from their horses, or rather tumbled off, when near Yeoman's saloon, Main Street. One of them (Anderson) was picked up insensible and carried into the above-named saloon. . . . There was an immense throng of men on the street at the time, and no sooner had the unfortunate equestrian been taken into the saloon than a great rush of the excited crowd followed. The building was soon thronged to the utmost capacity, probably not less than one hundred men having gained admittance, when suddenly and with terrible crash the underpinning of the house gave way, precipitating the whole of the already excited crowd a distance of eight or ten feet into the basement story of the building — lager beer barrels, stoves, chairs, bottles; in fact, everything used in the saloon following suit. The excitement which ensued can better be imagined than described. From the street and neighboring buildings hundreds of excited people flocked to the scene of the catastrophe, while amid the debris of the general wreck below, the cries of the panic-stricken throng were distinctly audible, and carried terror to every heart within hearing. . . . The woman who had been carried into the building unconscious was thoroughly awakened and was among the first to escape.[133]

Both miners and merchants were in constant danger of injury or death resulting from irresponsible blasting operations. In Treasure City, dynamite was used to grade the streets and level lots for buildings. The press continued to plead for some ingenious person to invent a process whereby this could be accomplished without so many explosions being fired off in surface rock. Considerable damage to property continued, and the escape of human life in a few cases appeared miraculous. By May, 1869, officials of Treasure City had adopted an ordinance making it illegal to set off blasts in the city limits and within two weeks a half-dozen miners were arrested for violating the law.[134]

No control over blasting operations in the mines on Treasure Hill appeared feasible. One observer noted, "There is a constant explosion of blasts in a thousand prospecting shafts, the air is filled with flying rock from the 'shots,' as they are called, and one might well fancy he listened to the roar of a regular battle."[135] Serious injuries were legion. The case of John Murphy was typical of many. This miner had placed a charge of powder in a rocky hillside, but when the fuse was fired and no explosion followed, he attempted to drill it out. While doing so, notwithstanding water used in the hole, the powder exploded, severely lacerating his hands and face. Murphy would have lost his eyesight but for the aid of the local physician.[136] All people were not so fortunate. During the winter of 1868–69 one Negro worker lost his life as the result of the carelessness of his partner. Apparently the latter did not learn a lesson because in the springtime he blasted himself out of the mining world.[137]

In the predominantly male society, controversies were often settled by a resort to physical violence. Disagreements appear to have arisen most often among men who were mining partners. In Shermantown James Butler and Thomas Davison, who shared the same cabin, got into an argument in the Sun Burst saloon; Davison shot Butler through the left elbow, breaking the bone, and in turn had one of his toes cut off with an axe.[138] Following the fracas both men were arrested for shooting firearms in the city limits. Bad debts were the source of much ill will between men. A miner named Callahan called at the cabin of one Adams two miles below Shermantown and demanded the money owed him. In an ensuing scuffle Callahan inflicted six knife wounds on Adams, with a large-sized pocket knife, one of which entered his chest just below the heart. After placing Adams, a young man of twenty-four, on his bunk, Callahan went to town and gave himself up to the authorities.[139] Teamsters who struggled to get their cargoes through the crowded streets often fell to fighting. When two wagons locked wheels on Treasure City's Main Street, one driver alighted from his wagon and threw a large piece of chloride at the other. The assaulted party jumped from his wagon and inflicted a severe wound on his adversary by hitting the side of his head with a six-shooter, almost severing his ear.[140]

Street fighting reached an acute stage during the spring season. "Spring Fights Opened," announced the *White Pine News* on April 20, 1869. "Main Street presented rather an animated appearance yesterday in the way of knock-downs, there being no less than three in different portions of the city during the day." If a fight occurred on Saturday afternoon or Sunday, the saloons poured forth their contents, and a crowd of two thousand men gathered to watch the outcome of the struggle.[141] Fighting was by no means confined to the male element of the population. Hamilton's *Inland Empire* reported the following:

"Poker Game — On Sunday evening last a couple of disreputable females had a little set-to downtown, one using a slingshot and the other an iron poker. One made a bluff over the other's eye with the billy, and received a call from the poker, which did about the same amount of damage."[142]

Prostitutes were the cause of countless brawls. The newspapers repeatedly noted events like the following: "A man got on a spree Saturday night, went to a house of ill repute corner of Treasure and Dunn streets, and pitched into the nymphs after the most approved style of 'knock down and drag out.' He dragged two of the women into the street, knocked them down and trampled them in the mud. He was finally captured and locked up."[143] White Pine women proved themselves an equal match for the men on occasions. For example, "a gay and festive hombre who sails under the euphonious and perhaps significant soubriquet of 'Boxing Barney' while under the influence of 'last spike' whiskey called upon Madame Stelle — 'frail, fat, and forty' — to proclaim his long-cherished and profound affection and to bring her to terms. A few months earlier the two had been on the most intimate terms, but Boxer was now a discarded lover. 'The brandy-perfumed object of his affections' damned and denied him, so he opened hostilities by firing a shot into the ceiling. She retaliated with a shot mowing a swath of hair across the top of his head several inches in length. At this point the law arrived."[144]

Periodically a well-known citizen gained notoriety from "woman trouble."

> The Treasure City people, so we are informed, are greatly shocked at a bit of circumstantial evidence which was — a few mornings since — brought to light against one of their city fathers. It seems that this functionary is not married. Still, the Ethiopian who attends to his sleeping apartments, found, one morning last week, in the very *sanctum sanctorum* of this man's sleeping room, a hair pin, one of the most elaborate workmanship, and tipped with an opal ball. Hanging the key to the room on this mysterious pin, the "image of God in ebony" approached the city father — who was just entering a company of friends — and exclaimed, "You's de bad man, Mr. Mike; you's de bad man, sar; but here's yer key."[145]

On another occasion a gentleman of Treasure City was forced to "air his linen" as a result of getting into the wrong berth on retiring.

> . . . as no objection was made by the hostess, he concluded to remain until morning — the host being absent. Before daylight, however, his repose was disturbed by a rap on the door, which was followed by a demand for admittance, made in a stentorian and resolute voice. The request was denied by the lady of the

house, when the voice outside threatened to "shoot the door to pieces"! This brought about a crisis. The door was hesitatingly opened, and as one guest slid in the other pitched head long out through a back window and beat a hasty retreat across the snow, with a pair of boots and a bundle of ready made clothing under his arm.[146]

In Hamilton, a woman of the town was arrested in a Main Street brothel, and charged with being *non compos mentis*. Upon examination, local doctors concluded that the derangement of her mind was due solely to too free use of the "ardent," so she was allowed to go on her way rejoicing. "Only a few reptiles in her boots, that's all," commented the town's newspaper.[147]

The life of these professionals was so miserable the mining camp journalists abandoned their humorous reporting style in favor of pathos.

Of all the human wrecks caused by the maelstroms of vice and dissipation, there is none so sad to look upon as that of a young and beautiful woman. We saw such a one in Hamilton. She was young, perhaps not twenty, but intoxicated — beastly drunk. She had staggered in through the back door of a gambling den, and been ejected from the room into the back yard (an open lot), where she fell, amid the jeers and ribald jests of the rabble, the mass of whom were doubtless less refined and no more virtuous than even the degraded target of their vulgar wit. In this helpless, maudlin mood, she began singing, in sweet and plaintive tones, the very words above all others most suited to her case: "Once I was happy, but now I'm forlorn."[148]

None were treated as beastly as the Chinese women from the brothels. One prostitute tried to run away from her owner and hide in the hills, but she was finally captured and held prisoner. Living in the open, exposed to the elements during her brief period of freedom, she had frozen both feet. The flesh fell away from the bones before her master asked admission to the hospital for her and then both feet had to be amputated. Although the wounds healed rapidly, the patient courted death, refusing to take medicine or food. She was eventually returned to the home of her owner to pass into oblivion without a protest from society.[149]

Far more serious than the sporadic fighting of drunks over women and money were the attempts at homicide and premeditated, brutal murders. The first death by violence occurred in Treasure City the last week of February, 1869. Two men seated at a table in the Mammoth saloon started a quarrel, challenged each other to a fist fight, and headed for the door. The man in the forefront ascended the stairs to the street and turned as his opponent stepped on the stairs and fired upon him, the ball striking the right breast near the waist and coming out at the loins. The murderer fled to Shermantown where he purchased a horse and left the district. The town newspaper wanted no second such affair, "so disgraceful to frontier towns, and which have made them a dread and a terror to the timid of the older settled communities."[150]

The White Pine communities were shocked by a double murder in May, 1869. Near the top of a mountain overlooking Hamilton on the west, two mining partners living in the same cabin quarreled late at night and one was secretly murdered. The following day the murderer called upon two men in a neighboring cabin to ascertain that at least one of them would spend that afternoon and night in Hamilton purchasing supplies. That evening to hide the crime, the murderer prepared to burn the body of his partner on the ground outside the cabin. With ax and sledge he hacked, severing the head and cutting the legs and body into several pieces. While the murderer was thus engaged, the neighbor returned unexpectedly from Hamilton in time to see the burning and charred remains. Realizing his plight, the unwilling witness attempted to plunge down the hillside two hundred yards away, but was not fast enough to escape four bullets in his back. The partner of this unfortunate was awakened in his cabin by cries of "Murder" and "I am killed." Thinking he recognized the voice, the man dashed into Hamilton to check on the whereabouts of his friend and learned that he was supposed to have returned home several hours earlier. He notified the authorities who reached the scene of double murder after the culprit had fled. Officers followed the murderer to Elko, on to Salt Lake, and two hundred miles beyond, but finally lost the trail.[151]

The community continued to be shocked by periodic murders through the fall and winter of 1869. Before the year was out ten men had lost their lives.

Added to the difficulties in the mining camps were those in the vicinity of the stage stations on the road to White Pine. Five-mile House on the Elko road was notorious for shooting affrays; one observer suggested "it must be fighting whiskey they have in that vicinity, as either a shooting affair or row of some kind occurs there almost daily."[152] Near Twelve-mile House, on the Egan Canyon Road, a Mexican shot and killed a Peruvian by the name of "Joe" who was his partner and cabin mate. No one witnessed the murder, but the men had publicly threatened to kill each other. The Mexican left the country on horseback leading his partner's fully-packed horse. Two weeks later their cabin was opened by law. Joe's bloody shirt was found with two bullet holes through it, and neighboring Indians were assigned the job of hunting for the body.[153]

Many White Pine residents who escaped the violence and homicide were preoccupied with the protection of their property. "Watch out for thieves" was the constant cry during the boom seasons. Desperate or mischievous men introduced the practice of taking what they needed. Blankets were a specially prized possession and burglars were urged to be more considerate since these were as "necessary to the happiness of their owners on cold nights" as to those who had removed them.[154] The light-fingered fraternity plagued businessmen in particular. A store on the lower end of Main Street, adjoining Chloride Flat, was pilfered by a robber who entered an adjoining unoccupied tent and cut a hole in the canvas wall of the store. Reaching inside, he took two hundred cigars and several bottles of wine. Apparently frightened by the return of the proprietor, he fled in a hurry, leaving behind a pair of felt overshoes he had been wearing.[155]

Housebreaking became a weekly, if not daily, occurrence. In Hamilton, burglars entered a lodging house, obtained a trunk from the sleeping quarters of the proprietress, took it to a ravine nearby, cracked it open and secured nine hundred dollars in coin, a diamond ring, and a gold watch.[156] During the summer months of 1869 an active police force reduced the amount of thievery, but in spite of their vigilance, night prowlers became unusually active as winter approached. Cabins on the outlying flats were entirely at their mercy.[157] Protests were soon voiced in the local newspapers:

> A number of suspicious-looking characters, who floated on the troubled surface of last Winter have been hanging about the sidewalks and saloons for a week or two past. It is right they hang around, but what troubles people is, no rope is used. Lately, a number of burglaries have been committed. Cabins are broken into and robbed while the owners are at work or up in town. Tuesday night, Leet's cabin, on Bromide Flat, was entered, and a watch and some other articles taken. The manner of entry was breaking down the door.[158]

Another source of difficulty came from ore thieves. One man, discovered in the act of carrying away a sack of ore from the Hidden Treasure dump by one of the officers of the mine, was told he did not need so much. The filcher replied: "Yes, I do; I am going to send a mine down to San Francisco to sell, and I want a sack of good ore to send along in order to sell the mine."[159] Another thief who wanted silver did not waste time sacking someone's ore, but obtained a skeleton key and entered the assayer's office in Shermantown and made off with four bars of silver. The precious metal, belonging to the Eberhardt Company, was passed out through a window of the melting room and packed away on the back of a mule. The culprit was found and forced to return the loot.[160] Four months later the Stanford Mill at Eberhardt City was robbed of 2,000 ounces of crude bullion worth about $3,000. While the mill was operating and the machinery making a terrific din, the pilferer entered the retort room, opened a wooden chest and filled several bags.[161] The sheriff finally arrested five suspicious characters, one of whom made a full confession and pointed out the places where the bullion had been secreted.[162]

In addition to the endless petty theft throughout the mining camp, the stage was occasionally held up on the road between Elko and Hamilton. At one o'clock in the morning on June 10, just as the Wells, Fargo stage reached the sandy plateau on the summit of Pancake Mountain, seventeen miles north of Hamilton, three men stepped out into the road and ordered a halt. The driver, George Carlton, noting several shotguns and rifles pointed toward his head, handed over the treasure box. According to one account, "A lady was on the seat with the driver, who exhibited most undaunted courage, showing no fear, but, on the contrary, was rather disposed to give the highwaymen a curtain lecture for their display of villainy." One thief, gun in hand, remained at the head of the team while two others retired to a convenient clump of bushes and chopped open the box. According to waybills of the company, the highwaymen obtained only $220. Bitterly disappointed, the men returned to the stage to inquire about a second treasure chest. After some parleying, the passengers, three outside and nine inside, were ordered off the stage while the robbers unloaded both boots, opening every package supposed to contain anything valuable. The personal effects of the passengers were not molested. The highwaymen then made inquiries about the time of arrival of the next stages and the amount of treasure each would carry. Although only three men participated in the robbery, the passengers all agreed that there were more hiding in the bushes, some mentioning two, others three. All culprits were reported to have had on spurs, one was said to speak with the accent of a Chinaman, but was thought by others to be a Mexican. Throughout the ordeal the bandits had conducted themselves in a gentlemanly manner toward the passengers. All were closely masked and left no clue to their identity other than the fact that one was tall and sparse, a second very short, and the third heavyset and of medium height. Wells, Fargo agents immediately offered a reward of $1,000 for arrest and conviction of the robbers.[163]

A few weeks later the Wells, Fargo stage was stopped again at the same spot. As usual, the robbers were three in number, a long man, a short man, and a thick man, and they now operated with dispatch. One stepped in the road to order a halt, and simultaneously the other two sprang up on either side of the driver. One man shoved a shotgun through the window and told passengers to keep their seats, while a second demanded that the driver hand over the treasure box. Once the box hit the ground, the driver was told to drive on. The whole affair lasted only two minutes and was conducted in "a very gentlemanly and businesslike manner." The amount of money and valuables taken was not revealed by the company.[164]

Residents of the countryside surrounding the mining district were continuously victimized by horse and cattle thieves. An auction of a mule was stopped in Hamilton when a former owner recognized the animal and knew that the man to whom he had sold it near Shermantown had complained of its loss. The man in possession of the mule agreed to lead authorities to the place where he had obtained it. A posse rode out into Ruby Valley where they came upon a band of fifteen to twenty horses among which Hamilton residents found many of their mounts. No men appeared to be attending the herd, probably having fled upon sight of the posse, so the stolen animals were all recovered and driven back to town.[165]

A livery owner of Hamilton had an entire band of eighteen head of horses stolen from his ranch four miles north of Hamilton. Accompanied by a horse wrangler, the owner started in pursuit of his stock, striking the trail near his ranch and following it without difficulty through the mountainous country to the Egan Canyon Road, twenty miles distance. Here three thieves were overtaken with the stock. The ensuing fight resulted in the killing of one thief, the wounding of a second, and the capture of a third. The courage of the livery man and his wrangler received the hearty approval of all respectable citizenry as the proper course of action in dealing with horse thieves.[166] The *White Pine News* reported that the trouble had centered in a well-organized gang of bandits whose central rendezvous was in Folsom, California. The leaders had displayed consummate skill in management and discipline, making conviction of any member of the "ring" next to impossible.

A rendezvous is established in some town or in the country for "prospecting" purposes; the keeper is furnished with money to start a business, which he or they follow strictly, meantime keeping an eye on the condition of stores and safes, making a note of their contents, as well as upon the treasure shipments, travel and time of stages. Whenever anything is noticed, which it is thought will "pay," information is furnished headquarters, where men particularly adapted to the branch of robbery to be committed are selected to execute the plans suggested by the rendezvous. The head man, the absolute ruler of the organization, gets half the plunder, while the other half is divided equally between the informant or keeper of the rendezvous and those who execute the plans.[167]

When the leaders of this California organization were finally identified and imprisoned, the members of the organization throughout Nevada were left stranded. Some made feeble attempts to rob the Wells, Fargo stages, but most of the band turned to horse stealing.

By 1870 a new type of horse and cattle thief put in an appearance. These men who habitually committed depredations against White Pine ranchers were not vagabonds, or wandering scoundrels, but wealthy ranchers with homes and families. They built their own herds by raiding the ranchers near Hamilton, and sometimes drove stolen cattle to a hidden corral in the Snake Valley of Idaho where confederates from Colorado arrived to take the animals to the mining camps of the central Rockies. Three of these "speculative ranchers" resolved to rob a neighbor who was known to be ready to depart for southern Utah with several thousand dollars to buy cattle. Before he returned with the herd they expected to steal, the wife of one of the plotters gave warning. The plan was frustrated, but the poor woman was almost beaten to death by her husband before she came to town to tell her story to the local newspaper. Such rustling operations were well-known and the editor of the *News* inquired:

> Can not the officers keep an eye on these fellows? In the East this sort of gentry have short shrift. If these are not apprehended by the proper officers, after a few more offenses, the people in the valley will give us a taste of lynch law. Already they have notified two of these fellows that they are to be hung up for the next offense. Beside the loss of property these thieves have had such an immunity from justice as makes their example poisonous. They are seen to prosper, though known, unmolested.[168]

When nothing was done by spring, ranchmen east of Hamilton adopted a new method of handling cattle and horse thieves — Spencer rifles on sight. Ranchmen turned out in a body to pursue men who had taken a dozen head of cattle from a local rancher. They came upon the culprits encamped in Ruby Valley. Here the firing commenced and it was believed that one of the thieves was mortally wounded for "he threw up both hands, dropped his pistol, reeled for a moment and then staggered into the brush." Parties went out the following day to search for the body, but it had been removed by his confederates. All the stolen cattle were returned to their owners.[169]

At times the law enforcement officers appeared so helpless in coping with stage robbers and horse thieves that the press grew cynical and satirical.

> Really our robbers have grown to be a feature on the Elko road, and the story of their exploits takes us back to the days of Jack Sheppard and Red Ridinghood. . . . We must judge them as we do other legitimate robbers, by their success, and in view of the last three days exploits, we lift our hats to them.[170]

The editor of *The Inland Empire* thought community opinion was primarily responsible for the violence and law breaking.

> We pay too much deference to bullies and violent men. . . . We are but frontiersmen, and we all of us applaud, if we do not practice, the code of violence. . . . Our Courts are perfect, and if justice is sometimes outraged before them, it is because public opinion is not right, and because in the minds of the jury there is an influence stronger than the oath they take to decide according to the law and the facts. It is right that the law should be executed, but for the general laxity of morals and indifference to crimes committed here, we are all more or less to blame.[171]

Periodically there was the threat of a lynching. While the press agreed that the guilty be punished, they urged that the responsibility be left to legally-constituted authorities and not to the public.[172]

The police vigorously enforced the law, but the trial juries in the mining camps had a tendency to be lenient with wrongdoers. In a single week of June, twenty-five men were arrested: nine for sleeping on the sidewalk, eight for disturbing the peace, three for assault and battery, one for committing a nuisance, one for murder, and one for indecent exposure. Commented the town newspaper, "In justice, we must say we have never seen any city under better or more strict discipline than Hamilton, as regards the police department." However, when these twenty-five violators appeared in the justice's court, only three were committed to serve time, six were discharged, and the rest fined a total of $168.[173] This disposition of cases seems typical of most weeks in both Hamilton and Treasure City.[174]

With the pressure of increasing population, the police had more and more difficulty. Many culprits escaped and others broke jail by digging out under the foundations or by loosening and removing bricks in the side-wall.[175] By mid-August the county jail, about fourteen by nine feet, housed sixteen prisoners. A row of bunks on each side of the room reduced the standing area one-half; this crowded pen was provided with no light or ventilation. The local newspaper fumed, "To huddle sixteen men together in such a place as this, though criminals they be, is a positive shame and disgrace to any Christian people."

Inmates were reportedly eaten up with vermin, and only the kindness of fate had prevented an outbreak of pestilence. The editor thought it a wonder that they had not "sickened and died" and he demanded that the County Commissioners improve conditions immediately. Friends of six miners incarcerated in the county jail were urged, in the name of humanity, to raise money to pay their fines and obtain their release.[176]

In spite of periodic outbursts of enthusiasm when the press offered to "pit Hamilton against any city of its size on this or any other coast for the general good order and sobriety of its citizens," the reports of the grand jury indicated that crime went on unabated.[177] In March, 1870, bills of indictment were found:

> For murder in the first degree, two; assault to commit murder, one; grand larceny, eleven; forgery, two; dereliction of duty in office, two; embezzlement, two; obtaining money under false pretences, one; highway robbery, two; drawing and exhibiting deadly weapons, one; malfeasance in office, one; breaking jail, one; perjury, one.[178]

The *White Pine News* suggested that even this gloomy summary of conditions was a whitewashing of the real corruption and violence in the mining district, and then fell to quarreling with the rival Hamilton newspaper over the issue.[179] The uneasy society of the rough and tumble camps in the White Pine Mining District during the rush was the inevitable result of sudden wealth, the good luck of some, the despair of more, and an "easy come, easy go" philosophy.

RECREATION

The miner's recreation was limited, masculine, and crude. Drinking and gambling were as inseparable a part of his life in White Pine as elsewhere in the mining kingdom. Physical stamina was essential to social approval, whether demonstrated in a footrace, bare fist fight, or in a dance with a male partner in a saloon. Contests of physical strength, by residents or professionals, provided the miner with an opportunity to indulge his gambling instinct. The miner's delight was the theater stage. Wandering troupes of musicians, vaudeville and dramatic players arrived more often in the Nevada mining camps than elsewhere. However, the theater never gained the stability in Nevada, and particularly in White Pine, that was attained either in California or Colorado. Music in general, and nostalgic ballads in particular, always seemed to charm the miner. Only a few found pleasure in reading.

Most of the miners were young, vigorous individuals in search of a good time during their leisure hours. During the winter of 1868–69, the principal amusements of White Pine men were hunting for silver ore when the weather permitted, writing home to friends, and talking about incredible developments on the hill. The most prosperous social institution was the saloon where men gathered for drinks, cards, and conversation. Down by the Eberhardt Mine the proprietors of the Bank Exchange Saloon were among the most influential agents for procuring development capital. The road to their establishment was lined with "capitalists" wanting to see the paying mines before investing, and many paused to "take a horn." All mines looked richer after their call.[180]

"Those hungry and thirsty souls who flock daily into the neat, cool, and genteel establishment of Goodman & McCluskey, on Main Street," commented the *White Pine News,* "have surely often cast a glance on the large painting hanging in rather poor light." This picture, representing the reclining Venus being crowned by Cupid, at whose feet sits a love-sick troubadour giving vent to his enamored feelings with music on a mandolin, was a copy of an original by Guido Reni hanging in the Royal Picture Gallery of Dresden, Germany. The editor congratulated the proprietors on their "refined taste," but rather deplored "the fact that three bullet holes in different parts of the body have somewhat reduced the value of the painting." Those who considered the picture a "fast" one were advised to banish the thought, for "a more chaste work of art has rarely existed."[181] Life was seldom dull in the saloons. One newspaperman, who dropped by a Hamilton establishment where some young clerks had stopped to throw dice for a beverage, was greatly amused at one excited member of the party who, when his drink was set in front of him, "caught up the dice and, putting them into his cocktail, began to shake it furiously. Had not the liquid splashed on him, bringing him to his senses, the whole contents of the glass would have been poured on the counter."[182]

When spring came, miners engaged in all types of competitive sports. Most popular of all was the footrace. On May week-ends as many as fifteen hundred people gathered to watch a series of such events. The preliminary was usually run between two youths with $100 bet, followed by the main event for as much as $500, and a finale in which a representative of Hamilton would run against a man from Treasure City with a basket of champagne as the prize. "In addition to the wine, the fastness of the two towns was at stake." On one occasion the Hamiltonian's victory by one and one-half inches was explained by the fact that he had exercised his muscles for the last three years wading through sixteen-inch snow to get to Treasure City. Hamilton, of course, drank with the winner to celebrate the victory. Even the local editor remarked, "We did not see the race, but we drank. And thus closed the spring races."[183]

A miner by the name of Johnson laid claim to the championship. Each week he ran a one hundred-yard footrace with $1,000 a side as the prize; he finally agreed to run with his hands tied behind him in order to find competitors.[184] As summer progressed, more innovations were introduced. Johnson ran one hundred yards to Spanish John's fifty yards, the latter, however, carrying a man on his back. Johnson won again by running his one hundred yards in ten seconds. Following this engagement, an impromptu race between two mustang horses was promoted, with the loser buying whiskey for the crowd.[185] By fall, the stakes on footraces were running as high as $3,000.[186] Those who were not adept at running, often spent Sunday afternoons participating in jumping contests.[187] Periodically a grand shooting match was scheduled when the crack shots of the camp would compete for a prize of $500 or $600, victory going to the man who made the six best line shots. A turkey- and chicken-shooting match sometimes followed so that all who attended might have a chance to participate.[188]

In the spring of 1870, a baseball club was formed that claimed to be the heaviest ball club on the Pacific Coast, and perhaps in the world. The twenty-one members of this Fat Man's Baseball Club of White Pine had an aggregate weight of 4,856 pounds, each man contributing from 220 to 260 pounds to the total. By summer another group known as the Pogonips challenged New Club to a match, and the resulting score— Pogonips, 58 and New Club, 47 — showed that the teams were not too unevenly matched.[189] An athletic club, with twenty-five founders, was organized in Hamilton and a committee appointed to purchase the furniture and apparatus needed for a gymnasium. The group planned to rent a hall and send to San Francisco for the necessary rigging.[190]

For those who preferred to be spectators rather than participants at athletic events, prize fighting was the featured form of entertainment. Among the earliest arrivals of the winter of 1868–69 was Joe Coburn, noted pugilist. His announcement that he would give a sparring demonstration the following Sunday at the "stone saloon" in Hamilton created considerable excitement in the community.[191] The first White Pine prize fight was announced for Sunday afternoon, April 16, "somewhere between Treasure City and Hamilton." Edward Fitzgerald and Johnny Murphy, lightweights, were to fight for a purse of $500. Fitzgerald had fought thirteen battles in the eastern states and in the Montana mining camps and was "quite a boy yet." His second, Barney Duffy, had recently engaged in a fight in Mexico that had lasted six hours and thirty-five minutes resulting in the death of his antagonist. Murphy's second was to be Johnny Grady, champion of the lightweights, who had just finished an engagement in Virginia City.[192] The day before the fight, the press reported, "Both men are in good condition and 'eager for the fray' and we expect to see the mugs of each considerably enlarged before either will permit the sponge to be elevated."[193] At the appointed hour Fitzgerald failed to appear, and Murphy therefore claimed the lightweight championship of the state of Nevada. It was reported that "no fights occurred among the disappointed crowd, owing, perhaps to the fact that a driving snow storm prevailed at the time."[194] In an attempt to promote another contest, Joe Coburn offered $5,000 to $10,000 to back his Irish friend, Mike McCoole, against any man, the fight to take place within sixty days after signing the agreement. Joe was willing to stage the fight anywhere west of the Rocky Mountains, but he preferred a place "on the great desert" where the roughs could not interfere.[195] No challengers appeared.

The long-talked-of prize fight between Johnny Grady and Johnny McGlade, two Irishmen, aged twenty-six and thirty-four, came off on Sunday afternoon, August 15, with $2,000 at stake. There had been much speculation as to the site of the approaching event, but those promoting it sedulously refrained from naming the locality for fear the marshal might stop the bout. By word of mouth, the interested populace learned that the "secret" spot selected for the battle was about one mile northeast of Hamilton on the Egan Canyon Road. The tent of a traveling circus was placed on the ground and those wanting to witness the performance were required to pay three dollars. Three to four hundred spectators wagered approximately $1,000 on the outcome, with the odds even. Within the twenty-four foot ring, the contestants were to fight half minute rounds. After battling thirty-four rounds in two and one-half hours, McGlade, who seemed to be getting the better of his opponent, walked to the center of the ring and gave his hand to Grady. He remarked that his hands and wrists were spent and it was utterly useless for him to continue the fight, as every blow inflicted punished him more than it did his opponent. One reporter suggested, "Although Grady acted on the defensive a greater portion of the time, he was the most severely punished of the two. McGlade's injuries were principally confined to his hands and wrists, which were pretty badly used up, one finger of the right hand being entirely out of joint."[196]

Shermantown gained notoriety as a horse-racing center. One Frank Merrill had a grey mare that was the pride of the community, and each Sunday afternoon of the summer the Shermantown track was crowded for the races. The distance run was from 300 to 400 yards; the bets anywhere from $200 to $500. These events did not always end happily if the race was close.[197] If the judges failed to agree on the victor, the stakes were withdrawn. The following spring two promoters had improved a course on the Beachy Road to Elko about four miles north of Hamilton, and Sunday afternoon horse races were centered there.[198]

Old and young, male and female, looked forward to the arrival of the circus. In June came news that the Jim Miller and Company Circus with George Constable, a well-known clown, would arrive in White Pine.[199] The performers, sponsored by the railroad, arrived in Hamilton on July 3 and the following day the wagons and ponies put in an appearance. After a two-day stand in this town, the group moved up to Treasure City where a large audience was entertained by the "little trick pony" and the excellent tumblers in the troupe. As the crowd fell off, admission price was cut from $2.00 to $1.50, and then the group moved on to Shermantown.[200] A week later the circus disbanded. A few of the performers left for San Francisco. Patronage in White Pine had been good and the proprietors made money, but most of the employees," getting chloride on the brain," decided to quit show business and go in search of an Eberhardt. The press insisted there had been no "bust up" on account of financial difficulties. The circus tent was left in the "plaza" of Hamilton.[201]

Dancing was always a favorite pastime. Because of the dearth of ladies, men often devised impromptu dancing matches with their male companions in the saloons. One night in Hamilton, James Finnegan challenged Johnny Rowe to an endurance dancing contest. The two men bet their coins, coats, hats, boots, and shirts on the outcome. A fiddler was procured, all the spectators took another drink while the fiddle was "greased," and the dance began. The fiddler was exhausted before the dancers so it was agreed that each partner should retain his clothes.[202] A number of young men of Hamilton, looking forward to more refined dancing with ladies, organized a club to sponsor a series of hops during the winter season of 1869–70.[203] The first affair under the auspices of this group, the Young Men's Christian Association, occurred early in November at the Fireman's Hall.[204] Later in the month a grand ball was given at the new Courthouse in Hamilton. The arrangements committee provided carriages to bring the ladies from Shermantown and Treasure City, and the press commented, "From appearances we should judge that about all the able-bodied people in the country are to be in attendance."[205] The parties of the Young Men's Social Club were bigger and better the next season. The editor of *The Inland Empire* commented:

> Last evening, hearing dulcet music at the City Hall, we stopped in, and oh! ye gods, the sight of beauty, elegance and splendor that met our eyes — thirty-four couples of ladies and gentlemen on the floor waltzing. We became so intoxicated in the dance, that

even now we feel frenzied, like one intoxicated on ambrosial wine by goddess hands brewed and served. It is impossible to describe this incomparable hop. We would recommend all pleasure-seekers to join the Club.[206]

White Pine residents seeking other types of refined entertainment and recreation formed musical groups. The White Pine Brass and String Band, consisting of eight instruments, was practicing daily during June, 1869, in anticipation of a concert on Independence Day.[207] Ladies and gentlemen in Shermantown established the Silver Springs Glee Club specializing in "promenade concerts."[208] Without doubt the most ambitious musical undertaking of the community was the White Pine Philharmonic Society, through which "a number of fine voices, male and female, under the intelligent direction of Professor Plumhoff" approached a perfection which placed that society "on a footing with the best amateur musical societies in the country."[209] In March, 1870, the society gave a "grand vocal and instrumental concert" concluded with a dance. Shortly thereafter, the group decided to adopt a constitution, and when the question of "woman suffrage" came up the men decided to be generous.

> The debate and vote, taken together, afforded the finest exhibition of the subtlety, and even frailty of man we have ever seen. No ladies present. Only two married men. Singular as it might appear to some, both the married men opposed the amendment with vehemence. It did not strike us as being particularly strange, however; for we thought married men ought to know whether women were capable of exercising the suffrage. The married men opposed woman suffrage in debate; but they saved their top by voting right — and the vote in favor of it was unanimous.[210]

The second spring concert was received with mixed feelings.[211]

Pleasures and luxuries of life increased in direct ratio to the population. Ladies and gentlemen of Hamilton could patronize the Metropolitan Bathhouse on the corner of Main and Dunn streets and "luxuriate in fresh pure water until their hearts' content." A corps of first-class tonsorial artists were connected with the institution and gave special attention to the dressing of ladies' hair.[212] In Shermantown an ice cream parlor was opened in the rear of the Orofino Saloon opposite the theater to cater to those attending dramatic performances.[213]

White Piners always had an adequate amount of periodical literature to read. The Postoffice Bookstore in Hamilton carried *Harper's Monthly*, Frank Leslie's *Pleasant Hours, Punch, Scientific American, American Artisan, Harper's Weekly,* and the *Waverly Magazine.*[214] A few doors

from the First National Bank in Treasure City an indefatigable news-
dealer by the name of Robertson had such other periodicals and picto-
rials as *Godey's Lady Book, Atlantic Monthly, New York Ledger,
Chimney Corner, Sporting Times, Illustrated London News,* and the
New York Weekly.[215] A circulating library also was opened in the
Hamilton postoffice.[216]

White Pine society became more complicated and highly organized
during 1870. Several fraternal groups, including the Masons and the
Odd Fellows, incorporated in each of the three larger communities. In
Shermantown the Masons and Odd Fellows combined their financial
resources and built an impressive meeting hall.[217] Perhaps the most
social organization of all was the White Pine Hook and Ladder Company
No. One of Treasure City. When the truck, ladders, and other parapher-
nalia first arrived in Hamilton, the "fire laddies" assembled and marched
to the toll house, half-way point on the Hamilton-Treasure City road, to
escort the "machine" to the city on the hilltop. The ceremony installing
the equipment provided the excuse for a "gay day in the annals of
Treasure City."[218] Within two weeks, residents were being invited to buy
tickets to the Firemans' Benefit Ball to help pay for the equipment.
Soon each community had one or more firemen's organizations. At the
end of the year the Hyman Engine Company, named for the mayor of
Hamilton, staged a mammoth parade to celebrate the arrival of its new
engine. The procession included the White Pine Brass Band, a carriage
containing the City Fathers, followed by members of the Liberty Hose
Company No. 1, the Hamilton Hook and Ladder Company, and finally
the new Hyman Engine Company, all dressed in bright new uniforms.
They "marched and counter-marched through all the principal streets
of the city," and finally "many speeches were made and much wine
flowed." The group adjourned to the City Hall for the evening where
the honorary members gave a levee for the firefighters.[219]

National holidays, both of the United States and of the "home
country," provided an additional opportunity for the miners and their
families to celebrate. On the anniversary of the natal day of Ireland's
patron saint in 1869, a terrible snowstorm prevailed, and the Irishmen,
who were numerous in White Pine, were quieter than expected. Special
masses were observed, Madame Minna Bernard gave a concert of Irish
songs appropriate for the occasion, and that night two enthusiastic young
sons of Ireland paraded the streets with drum and cornet, regaling the
citizens with "Wearing of the Green" and other favorite Irish tunes.[220]
By the next season, the Irish were better organized. In Treasure City
a grand ball was staged with a first-class bar in the basement. Special
suppers were held in Hamilton's restaurants, and the White Pine Pioneer

Brass and String Band provided the music for a rollicking dance, the proceeds going to rebuild the community's Catholic Church destroyed by winter storms.[221]

National independence was observed with a three-day celebration in 1869. Starting in Shermantown on July 3, festivities included speeches by the City Fathers and the sending up of a balloon which unfortunately "caught fire" and disintegrated after a one hundred foot ascent. In Treasure City the following day there was a street procession, more orations, and poems read. The climax of the observance in "silver land" came in Hamilton on the fifth where the parade was bigger, and the orations longer.[222] Before the next annual celebration, a local editor suggested:

> As most of the scattering population of White Pine naturally point towards Hamilton on festive occasions, we submit to our neighboring towns that it would be well to concentrate all their influence upon the celebration of the Day at the county seat, instead of wasting their liberality in separate and perhaps tame demonstrations in each burg. It would be like making several bites of a cherry to hold more than one celebration.[223]

His viewpoint prevailed and the festivities were held in Hamilton.

Washington's birthday was observed by a parade of the combined fire companies, all dressed in their uniforms of red shirts and black pants. That evening grand concerts and balls were held in all three towns. In Shermantown the affair was under the auspices of the Silver Springs Glee Club, and in Treasure City was sponsored by the Liberty Hose Company. The fire company was commended for exercising due vigilance at the door "from which all improper characters were excluded and the assemblage rendered select."[224]

Thanksgiving was celebrated with the usual feasting and drinking, but the religious services held in both Hamilton and Treasure City were poorly attended.[225] In addition to private dinners, the public houses treated their patrons to "something extra in the way of edibles." Balls were held in both places with the firemen of Treasure City again in charge in that community. The affair was pleasant and befitting the occasion, so the newspaper commented, "Our fire boys are no roughs, and are none the less prompt and efficient workers when the hour of danger is on. They are gentlemen in a parlor or a ballroom, but lightning at a fire or a fight."[226] As the Christmas season approached, the Scandinavians called a meeting of the people of "the Old North Country" to meet in a Treasure City bakery to plan a social reunion and observe the holiday according to the joyous customs of their youth. The upshot of this meeting was the formation of a Scandinavian Society for social and charitable purposes.[227] Group singing of carols and hymns around the Community Christmas trees at the Courthouse in Hamilton and at Broker's Hall in Treasure City, followed by a turkey dinner for the entire town, were the highlights of the Christmas observance in White Pine.[228]

The most unique and widely-publicized event in White Pine's heyday, providing for the gambling instinct of the populace as well as for their entertainment, was Wright and Wray's "great gift enterprise," a public lottery. These enterprising gentlemen arrived in Hamilton with a supply of gold watches, pen knives, and jewelry that were placed on display in a local saloon. Miners were encouraged to buy tickets entitling them to a chance on a grand prize of $5,000 or to lesser prizes of jewelry valued at $200 or $300. In the beginning tickets were raffled off for "four bits" each, but later rose in value to two dollars. Wright was described as "one of the oldest and most reliable business men in the states" who would never be guilty of a "shennanigan."[229] Tickets were sold for two months from late April to late June, 1869, and the drawing for prizes was scheduled for July 4 at the Wells, Fargo building in Treasure City. As the day approached, excitement mounted and the prospective winner was notified by the press that he was expected to treat the whole crowd.[230] The promoters postponed the drawing until August, hoping to increase their profits. On the appointed day a committee of eight citizens was elected to superintend the drawing; two little girls were chosen to draw the numbers. The winners of the first prize of $5,000 and the second, a cluster of diamonds, were not present. Approximately five hundred gifts of lesser value were distributed. One miner received a neat silver fruit knife, but returned it almost immediately saying, "I tried to whittle with it, but the dod rotted thing wouldn't cut worth a dern!"[231] A few days later the newspapers announced that an Elko resident won the diamonds and a San Francisco man had the lucky ticket for $5,000. The Californian sent each of the "misses" who officiated at the drawing, a $125 gold watch and "stood treat" liberally at the saloons in Hamilton. The following day Wright and his associates left the mining district.[232] Three weeks later a prospector came down from the hills with a wad of prize tickets in his pocket and, in the presence of the editor of the White Pine News, he pulled out one that had the lucky number for $5,000 on it. The newspaperman thought it was legitimate, but the promoters' representatives insisted it was counterfeit. Rumors began to circulate that the Elko and San Francisco winners had been "plants" working with the raffle sponsors to defraud the public. Although the press thought Wright had a business reputation to think of, no amount of pressure brought him back to White Pine.[233]

No organization or event succeeded in upsetting the social primacy of the saloon. As competition for the drinking trade increased, some establishments introduced added attractions, foremost of which was draw poker.[234] In 1869, there were ten or twelve gambling tables licensed under the law of Nevada in the three larger towns of White Pine District.[235] A few saloons sponsored billiard matches.[236] Others constructed shooting galleries adjoining their bars to entice customers.[237] The greatest challenge to the business of the saloon keeper in the mining community was the theater, and when a good troupe arrived in town the bartender usually padlocked the door and went off to the performance.

THE THEATER

The theater was an ever-present and popular institution on the American frontier. Traveling musicians, acrobats, and dramatic artists with any talent always managed to draw a sizable audience. Nowhere in the West was the stage as vital and well-patronized as in the mining camps. In these predominantly male communities, the female entertainer was cherished and lionized. Without doubt, migratory players, some good and others mediocre, exploited the market to the fullest. In many camps only the bawdy proved popular. Elsewhere there was emphasis upon the legitimate theater as the first attempt at cultural refinement, particularly when family groups reached the mining district. The single men responded enthusiastically if there was enough romanticism and sentiment to provide an escape from the harsh reality of their daily existence. Repeated efforts were made to determine the changing tastes of the community by offering both the bawdy and the cultural.

The first theater in the mining district, located in Treasure City, was known as the White Pine Theater. Here, in March, 1869, the "young and talented" Pixley Sisters, Minnie and Annie, were singing and giving dramatic readings for large and appreciative audiences.[238] In preparation for the arrival of Strasser and Levy's California Minstrel two weeks later, the entertainment hall was redecorated and additional seats installed. On opening night the theater was filled to overflowing. The Pixley Sisters were not ousted, however. "Miss Annie Pixley rendered a ballad 'We Parted by the River Side,' in a sweet and plaintive voice in admiral [sic] keeping with the sentiment of the song, and was complimented by an encore and rapturous applause."[239]

Theater business became so profitable that a second house, Rain's Hall, opened on Main Street. Ten talented musicians, male and female, known as the McGinley Minstrels and Variety Troupe, were the initial attraction and "the lovers of good vocal and instrumental music, and of chaste, but genuine wit and humor" were advised "to avail themselves of this opportunity to enjoy an intellectual feast."[240] The premiere was a howling success. To meet this unexpected competition, proprietors of the White Pine Theater rendered the seats in their establishment "secure and comfortable," added a piano to the orchestra, and employed a new variety troupe, among whom was Charles Evans, the celebrated vocalist.[241] The local newspaper delighted in developments:

> The theater was well filled last night to hear the charming vocal and instrumental music to which this troupe treat their audiences each evening. The orchestra is full and complete, and there are a number of excellent singers and dancers in the company. An entire change of programme this evening, and no better place can be found to while away an hour pleasantly than at the White Pine Theater.[242]

Down the street Rain's was drawing only a fair house. An elderly tamborinist accompanied a "comic darkey" who "did 'bones' in superior style," but the leading attraction was Miss Sarah who played on her banjo and sang for the miners.[243]

In Hamilton, Blaisdell's Theatre Comique or Melodeon was crowded nightly in the spring of 1869. "The Jig dancing of Miss Josephine, the character songs and dances of Miss Clara, the splendid voice of Add Weaver, the clog dancing of Sam Rickey, and the polished acting of W. H. Vincent . . . keep the audience in a roar, and send the patrons of the house home each night better satisfied with themselves and the world than when they went."[244] For all this, admission was only one dollar, unless one wanted a front seat at a dollar and a half. The Blaisdell Brothers, theatrical managers and Swiss bell ringers, started construction of a new theater building early in April. This structure on Main Street, seventy by thirty feet, exclusive of a dressing room thirty by ten feet, was expected to be finished within two weeks. Meanwhile, the old institution continued to thrive. Crowds were summoned to the nightly performance by Blaisdell's bell ringing and, once inside, listened to a favorite cornet player until the dancing on the stage began.[245] When the new theater was completed it presented "quite a palatial appearance," and Blaisdell announced he had leased the property to Mrs. Rand who planned to "inaugurate the legitimate drama in White Pine." Mrs. Rand expected to present to the public her daughters, Miss Olivia and Miss Rosa, advertised as talented actresses and vocalists, supported by an excellent stock company.[246] Before opening night, the editor of *The Inland Empire* admitted, "We took a peep through a knot hole and saw some of the fine scenes now being painted. We do not advise everybody to go — we don't want to see more than 1,000 present."[247] Throughout the month of May the Hamilton Opera House, as it was now designated, was filled almost every night and performances were well received. When attendance slacked off, the public was upbraided for lack of support — "The performance last evening at the Opera House was most excellent. The attendance was not as large as it should have been, yet each performer played with the same éclat as though the house was filled to overflowing. . . . Go everybody."[248] The town's favorite cornet player was dropped from the orchestra because of flagging door receipts and the press complained, "We are sorry that Mr. Rippingham toots his horn no more at the Opera House. Cutting him from the orchestra is like taking the A string off a violin. It leaves the thing incomplete."[249] During the unseasonal cold of spring, the Opera House suspended performances so that a new stove could be installed.[250] Within a few weeks it was closed for the summer.

Traveling entertainers usually paid a call on Shermantown for a week's performances. In fact, the McGinley troupe selected this town for its White Pine premiere and established its reputation as a successful theatrical company before moving to Treasure City.[251] The Pixley Sisters, following their success in Treasure City, moved to Shermantown and became proprietors of a theater in that community.[252] One newspaper correspondent expressed concern when he wrote, "Who will say that Woman's Rights are not fast being established throughout this State, when the only two theatrical companies in the county of White Pine are under the management of women?"[253] After a two weeks run, the Pixley Sisters relinquished their theater to Tony Ward who attempted to introduce legitimate drama in the milling center. The effort was not an outstanding success.[254] Attendance dropped off because some of the plays, like "Black-Eyed Susan," did not appeal to the audience, but the theater was packed for a performance of "Cross of Gold" when the evening's bill also included a laughable farce.[255]

Mrs. Rand elected to sell her interest in the Hamilton Opera House to two theatrical promoters, Garrett and George, who promised to bring in a melodeon group with twenty performers, both men and women, in addition to a dramatic group. A group of "leading citizens" of Hamilton sponsored a benefit for Mrs. Rand who "for the last month has been catering to the desires of the people of White Pine." She announced her intentions of leaving on a tour of California.[256] Before going, Mrs. Rand herself put on a benefit performance for the town's pioneer theatrical man, Blaisdell, in appreciation of his construction of the Opera House she had purchased. Residents flocked to see this final theatrical performance knowing that the emphasis in the future was to be on the

variety show.[257] Instead of leaving for California, Mrs. Rand invested
some of her profits in a short-term lease on the White Pine Theater in
Treasure City, vacated by the Pixley Sisters, and moved her entertainers
there. The play selected for the opening performance was "the beautiful
and highly sensational drama, in five acts, entitled the 'Pearl of Savoy,
or the Mother's Prayer'."[258] Success followed success. First, "The Married Rake" and then "Maid of the Milking Pail" and the thrilling nautical
drama "Ben Bolt." The Rand Sisters were indeed charmers.[259] From
the successful run in Treasure City, they moved to Shermantown for
a few performances.[260]

The Rand Sisters were such favorites that all attempts at legitimate
drama following their departure were unsuccessful. Two promoters re-
conditioned a building along Shermantown's Main Street attempting to
provide a theater for a dramatic group, but the venture failed.[261] Shermantown opened a new Silver Springs Theater in midsummer that catered
to "gentlemen only" and provided melodeon entertainment. The enterprising promoters were bitterly criticized for second-rate performances:

> Our citizens were most egregiously bilked last night by a company
> of fiftieth rate melodeon performers under the name of the
> "Hamilton Opera House Melodeon Troupe." Although the house
> was a full one, it is impossible to find a man in town to-day who
> will acknowledge that he "bought any of it." We advise the troupe
> to "travel" for they can never draw a ten-dollar house in this
> town again.[262]

In Hamilton the management at the Opera House inaugurated the
practice of having "ladies' night" twice a week at which time only
"chaste" drama was presented, but the size of the audiences apparently
did not increase.[263] On the other hand, Emma Forrestall who was known
for her wonderful contortions and gymnastic feats arrived at the end
of June and the men of White Pine packed the Opera House to watch
her hold an anvil on her breast while a red-hot horseshoe was forged
and completed.[264]

On occasions, White Pine theatergoers witnessed a brawl between
miners arguing over the relative merits of the female performers on
stage. Life and limb were also placed in jeopardy at times by attempts
to provide realistic stage effects. Hamilton's *Inland Empire* reported on
one of these "dangerous experiments":

> The firing of a miniature cannon at the Opera House came near
> resulting disastrously last night. The wadding struck a hanging
> lamp, lighted with kerosene oil, and knocked it down. There was
> at once a stampede from the theater, men rushing over and tread-
> ing upon others. Fortunately the falling lamp did not break. Had

it done so the theatre would have been consumed by fire beyond doubt, and where the conflagration would have ended is left to conjecture. As it was, the wadding from the cotton dropped sparks of fire to the floor for several minutes after the bulk of it had been carried to the street, where it smouldered for more than five minutes. Better give up firing of that cannon for theatrical effect. The safety of human lives and the protection of property from fire require it.[265]

The New Melodeon in time replaced the Opera House as the center of Hamilton's theatrical entertainment and the level of performances deteriorated rapidly. The "Martinettis," a group of tumblers, performed "truly wonderful" athletic feats in mid-air for several weeks. They were followed by a local-talent minstrel show, known as the Star Minstrel Troupe, who put on such virtuous entertainment that the ladies could be present on the front rows.[266] Shortly thereafter a "previously unannounced" matinee took place that drew a large and appreciative audience. "It was pugilistic entertainment, in the course of which one of the proprietors of the establishment had a terrible head put on him. A refusal or inability to pay the salaries of the performers brought about the difficulty." The local newspaper announced, "We believe the show has 'busted'."[267]

In September, the Pixley Sisters announced they were leaving White Pine to make a professional tour of the towns along the Central Pacific Railroad. These "good actresses and perfect ladies" had been the greatest favorites that ever appeared on the boards of White Pine theaters and their departure marked the close of the pioneer period of the theater.[268]

Plans were announced in April, 1870, for the construction of the Hamilton Atheneum to be "the largest hall in the county" and "one of the most elegant theatrical palaces in the State." The management revealed that the first week's performances, including dedication night, should be devoted to the entertainment of ladies and their families with nothing indelicate permitted. When this place of amusement was thrown open to the public, there were seats to accommodate five hundred spectators. Private boxes had been handsomely arranged in the orchestra, and the "catacoustic" qualities were outstanding. Most impressive of all was the scenic decoration around the stage, including "a street scene in St. Petersburg, looking from the bend of the Neva towards the Emperor's Winter Palace" and "a wood scene, representing an everglade of Florida, with pampas and savannas interwoven with steppes." Amidst these surroundings, a traveling dramatic group presented Shakespeare's great tragedy, "Richard III."[269] The Atheneum remained the center for Hamilton's entertainment for several years. Alexander Montarg, the celebrated one-handed violinist, the beautiful Circassian Lady, and "other prodigies" performed here.[270] Meanwhile, Treasure City reverted its attention to dancing girls. Mlle. Kate Kasper, "the celebrated danseuse," and her troupe of ballet girls drew crowded houses in the camp on the hilltop.[271]

THE CHURCH, THE SCHOOL, AND HUMANITARIANISM

Men outnumbered women in all frontier communities and in the mining camps this inbalance was accentuated. Ladies were almost non-existent and their absence was of primary importance in shaping social attitudes and responsibility. However, even during the boom years there were a significant number of married women in the Treasure Hill area, and many of the mine and mill workers were family men who had wives and children to consider. White Pine miners, hardened individualists who paid little attention to community affairs unless their personal interests were threatened, neglected such important institutions as the church and school, but their negligence did not pass unnoticed and without protest. At the peak of population, however, society was not only predominently masculine, it was cosmopolitan. In this respect, White Pine was typical: there were rich capitalists, poor wage-earners, hapless prospectors, along with drunken loafers, rogues and thieves. The "yonder-siders" from California mixed with newcomers from the Midwest and many more men who had come from Ireland, Cornwall, and Germany. As usual, white men shunned and persecuted the Chinese, the Negro, and the Mexican in hopes of destroying their economic and social position. At the same time they were notoriously generous in aiding the unfortunates and underprivileged. The social attitudes and actions of the miners on Treasure Hill, for the most part, appear to have been typical of their counterparts in other mining towns in the western states and territories.

"We are a population of twelve hundred souls and are without a church," wrote the Shermantown editor in July, 1869. "Once in a while a stray preacher drops in among us and preaches to us. Why can we not have regular preaching? Why not have the good word regularly taught to our people? All need it, and at least we would appear more civilized and it might make our town appear more like a city in a Christian land."[272] For many months religious services in all White Pine communities had been conducted with itinerant ministers. In the midst of winter, Reverend T. H. McGrath, presiding elder of the Sierra district of the Methodist Church, had started preaching at the Court House in Treasure City each Sunday afternoon and, weather permitting, he held services that evening in the log storehouse in Silver City.[273]

The residents of Austin, Nevada, insisted that their religious arrangements had been interfered with "in the most damaging and damnable manner, by the demoralizing, all-debauching spirit of White Pine." The Methodist minister of the town, W. I. Nichols, familiarly known by the name "Big Aleck" had been "taken by the devil to the summit of Treasure Hill."

> There, above the clouds and the regions of the Pogonip, did the devil tempt Aleck with visions of rich chloride, silver bricks, and all the worldly allurements attainable with the possession of unlimited "soap." And Aleck fell. His flesh — of which he always had too much for a minister — was weak and Beelzebub had an easy victory over him.

Were it not for the kindness of a minister of another denomination, the new $40,000 Methodist Church of Austin would have been as silent as a tomb on Sundays. Moreover, Austinites thought their Catholic Church would have been finished with the $3,000 raised at a community fair if Father Monteverde had not gone off to White Pine to build another church.[274] By summer, both Catholic and Protestant services were regularly observed each week in Hamilton and Treasure City, but the number and influence of the congregations was not great. A movement was started early in 1870 to get Hamilton stores to close on Sundays; only one merchant agreed to shut his door at noon. The local editor thought White Pine could not take her place among civilized and enlightened communities until Sunday was recognized as a day of rest. "Every man in business will do just as much paying business if he closes his store or workshop on the Sabbath as if he kept open the seven days. If he loses some he will gain others."[275]

———•—•———

During the first year of the big boom, White Pine's only schools were conducted by private teachers in their homes, with students admitted on a subscription basis. When the first private school opened in Hamilton, the local newspaper suggested that the town's authorities should "extend public aid to the lady who has the pluck to start on her own means."[276] This plea fell on deaf ears, and when the school collapsed from lack of support the journalist penned an editorial:

> It strikes us that there exists a great want of public spirit and liberality among the principal landed proprietors of this city when they fail to set apart appropriate lots for school purposes. In all new American communities the very first step taken is to secure means of education for our children. . . . We have children here

> who are not receiving the benefits of our educational system for
> the want of proper sites. We have also men, who ought to see
> that in their nonage they shall not be deprived of them by indi-
> vidual rapacity.[277]

The crusade in behalf of education continued.

> Although this county has at the present time not less than 10,000
> of population, yet we believe there has not been a single school
> established in the county. It certainly is not for the want of pupils,
> for our streets are full of children who are being educated every
> day, not in useful knowledge, but in all the vices natural to a
> new mining camp. . . . White Pine is a fast country, and the
> majority of those coming here do so for the purpose of making
> money and no other, yet there are many families among us that
> have children, who, in order to make good law-abiding citizens
> should be educated, [but who] are receiving none of those advan-
> tages which are calculated to elevate them above the aborigines
> of the country.[278]

The State Superintendent of Instruction arrived in White Pine in
June, 1869, to meet with parents interested in the establishment of
public schools in each of the towns. State law provided that no appro-
priation could be made until a school had been maintained for six
months, so initial funds had to be raised by subscription or local taxes.
Citizens of Treasure Hill voted to provide the money by subscription,
and a collection committee was appointed. The town had eighty children
of school age, and $2,000 was the minimum amount needed to purchase
a lot and erect a substantial building.[279] The City Fathers of Shermantown donated a lot for that town's school, lumber merchants contributed
the lumber needed for construction, and businessmen were called upon
to subscribe funds for labor and furnishings.[280] To help finance the
public school system, "education committees" sponsored a series of
public lectures. Professor Israel Diehl spoke in each of the mining communities on the "Bible Lands" where he had toured.[281] In support of
the same cause, Rossiter W. Raymond, a geologist examining the country
for the United States government, talked on the mining industry of
Nevada in Broker's Hall, Treasure City. Raymond, a graduate of the
Freiberg School of Mines, was advertised as "one of the most interesting
and instructive speakers living."[282]

Hamilton conducted a school census in the summer of 1869 and
learned that there were ninety-five white boys and ninety-six white girls
between the ages of six and eighteen in the town. There were thirty-
three children under six years of age. In addition, twenty-three Indian
children and five Orientals of school age were registered. In all, Hamilton had 255 young people under twenty-one years of age.[283] Following

the same procedures adopted in Treasure City and Shermantown, a public school was organized and a schoolroom rented just off Main Street where there was enough space for a playground. The new principal, Henry Haven, who had recently arrived from California, urged White Piners to "take the youngsters out of the streets and send them to school."[284] When the school term opened early in August, only thirty-five would-be scholars appeared.[285] A mother from Treasure City was the first teacher, but "school marms" from Massachusetts, Vermont, and Pike County, Missouri had filed letters of application with B. H. Hereford, Clerk of the Board of Trustees. The town's newspaper had some fun at his expense. "Now, we sympathize with friend Hereford, for we know that his acquaintance with sage-brush (grass) widows in White Pine is very extensive and should an avalanche of schoolmarms, as appears imminent, come down on him, as sure as we are a sinner, he's a goner; lost forever in crinoline."[286]

In September a city election was held in Hamilton to determine if a school fund of $3,000 should be raised by a special tax on property. The money was to construct and furnish a school that could be used in the winter months. Otherwise, the term would continue to be limited to three months, from the first of August to the first of November.[287] The school tax was approved by a large majority and early in November a contract was let to build a new schoolhouse, forty-five by twenty-five feet, divided into two rooms to accommodate forty-two pupils in each. By the month's close, this building, described as "one of the most comfortable and best appointed in the State" was completed, and the public school term began. "With two such accomplished teachers as Miss Baldwin and Miss Hilton to direct the young, we can safely say that the public schools of this city are in a flourishing condition," remarked the editor of *The Inland Empire*.[288] Soon the public schools of White Pine were sponsoring "exhibitions" by the pupils in the form of group singing and recitations. The public was asked to pay a fee for these events to raise funds for school equipment such as a new bell.[289]

By spring, 1870, Hamilton was in need of another schoolhouse because the old one was filled to capacity and the town was full of children who did not attend. Since there was not sufficient coin in the School Fund for the construction, the press urged public-minded citizens to take the matter in hand and raise necessary funds by voluntary contributions.[290] Within a few years, school finance was systematized. According to the state-wide school census of 1876, White Pine County had 272 children between six and eighteen years of age and was thus entitled to $1,017 from the annual state appropriation.[291]

Several tragic incidents in the community occurred during 1869 and 1870 that touched the heartstrings of miners and caused them to unite in aiding those in distress. A family by the name of Woodruff — man, wife, and son of sixteen years — left Virginia City headed for White Pine in April, 1869. When they had journeyed as far as Egan Canyon, the old man took sick and died. With heavy hearts the woman and boy left the graveside by the road and finally reached Shermantown. They were destitute. A fund was raised by subscription, a lot purchased, and with volunteer labor a comfortable home was built for the newcomers. The Treasure City newspaper reporting the story thought "the action of the citizens of our neighboring town in this matter is certainly most commendable, and entitles them to the highest mead of praise."[292] No further comment was deemed necessary.

In July, 1869, W. P. Martin, the proprietor of a hotel and road station six miles west of Hamilton on the Egan Road, commonly known as Six-Mile House, was assassinated by robbers planning to hold up the stage.[293] Martin left his wife with a family of seven young children one of whom was blind. Late in August the press called the attention of the public to an unfortunate, footless boy who had been hobbling about the streets of Hamilton on his knees. He was Martin's oldest boy, age ten, whose lower legs had been cut off by a mowing machine in the Reese River Valley several years before. The lad was assuming his duty as eldest son to help support the family by setting up a fruit stand between the office of the Pacific Union Express and a local saloon. Not wishing to ask alms, he hoped for patronage.[294] During the winter, Mrs. Martin, broken by her grief and cares, died. Immediately a citizens' committee assumed responsibility for housing the orphans. Residents organized a grand raffle of goods made by the men and women of the town to raise funds for their subsistence.[295] In February, an old grandfather arrived from Kentucky to look after the children, but he too was poor and unable to work effectively for them. The public again arose to the occasion and sponsored a series of benefit lectures in the county court house.[296] Early in the spring, the community decided that for the best interests of the children they should be sent back to a Kentucky kinswoman rather than be raised unsupervised in a mining camp. Interested parties went to work to acquire transportation funds. In Treasure City, a benefit performance was held at the Opera House, including family entertainment by a brass band, singers, and comedians. On the day before this event two young ladies visited most of the business houses in Treasure City selling tickets. The editor of the News commented, "There was something touching and beautiful in the scene yesterday — two young girls, themselves as 'pure as the beautiful snow' that settled in fleecy clouds like a benediction upon their devoted heads, wending their way from door to door, through a storm, to ask of affluence a pittance for the relief of the orphans."[297] Proceeds were $175. The lecture fund brought the total to $240, and the Martin Orphan Committee had plans to raffle real estate that had been donated.[298] Miners who won the several pieces of property in this raffle elected to auction them off again to the highest bidder and collected an additional $825. The committee finally raised $1,928 and the press rejoiced that through "the large-hearted liberality of our citizens, these unfortunate little ones are now pretty well provided for."[299] A member of the committee, James Meteer, agreed to escort the seven children back to Kentucky, and the county recorder circulated a petition among the White Piners asking the Central Pacific Railroad to transport them without charge.[300] Word was soon received that various railroads had provided free transportation all the way to the home of their maternal grandfather in Stanford, Kentucky, and the funds raised for their aid were handed over to him intact.[301]

In the midst of winter two miners, losing their way along the trail between Hamilton and Treasure City in a severe storm, remained outside all night; one of them James Donaghue, had his hands, ears and feet frozen. Doctors insisted both feet had to be amputated, and the miner's wife, wanting to take him to San Francisco for the operation, offered her gold watch and a chain for raffle to raise money. Miners of Treasure Hill banded together and each agreed to give four dollars, a day's wage, for his relief. They raised $950. The Hamilton paper commented, "What a contribution from a town that has been isolated all through the winter!" Hamilton's own citizens added another $250. Friends of the unfortunate man finally convinced his wife that the amputation should take place at Treasure City rather than risk a trip to San Francisco. Both feet were cut off at the instep, the heels and ankles left so that, when he recovered, he would have club feet but be able to walk. Only the thumb and index finger of the right hand were saved and only the thumb of the left.[302] While the Donaghue relief committee was in Shermantown explaining their cause to a crowd of men in a saloon, a little boy who peddled apples in the town came in, listened attentively to the story until it was finished, and then walked up and handed one of the committeemen a quarter. The man refused the coin, but the lad insisted upon it, saying that it was little and all that he could spare, yet he wanted to give something for the good purpose.[303]

The White Pine community, full of such generosity for the oppressed, showed, on the other hand, little consideration for the minority racial groups among them. The few southern Negroes in camp were confined

to domestic service and menial tasks, and their antics made the subject
for jokes and humorous stories.[304] The Chinese served as cooks, laun-
drymen, and were literally hewers of wood and carriers of water. Even
so, their enterprise was respected. One reporter noted that, looking eastward from the top of Treasure Hill, a little green spot could be seen
along the slope of the mountain enclosed with a rough fence. Identified
as a vegetable patch appearing like an oasis in the desert, this property
was owned by "a couple of Celestials, who are said to be doing well."[305]
Although respected for his hard work, the Chinaman was expected to
work for less wages than the white man and he, like the Negro, was
the source of amusement and the subject for ridicule. One incident will
suffice to illustrate. An unfortunate washerman, Ah Qui, who had been
charged with technical assault on an Oriental beauty, was charged a
fine of ten dollars by the court and given a week to raise the money.
He worked all week for the best-dressed men in town — lawyers, speculators, printers — but had been "put off with jawbone" instead of money.
With tears in his eyes he appealed to the newspaper editor for aid, who
in turn confided in his readers:

> It is a hard case, and we felt for poor John, who will doubtless
> have to pine the fine away at the rate of $2 a day, in a loathsome
> dungeon. He tried to borrow a five from us, and we regret to say
> it was after *our* banking hours. We felt for him; but we didn't
> feel to the extent of half an eagle.[306]

When news of the outbreak of the Franco-German War of 1870
reached the mining camps, Americans displayed little understanding
of the conversational excitement among the Europeans gathered before
the newspaper bulletin boards. Soon after a French saloon keeper by
the name of Lazard raised the fleur-de-lis over the door of his establishment, the *News* commented, "Our citizens have already commenced
taking stock in the war between France and Prussia. Lazard was the
first in the field. Others are in for Prussia — but we don't suppose it
would pay either party to send a recruiting officer over here."[307] A few
nights later mischievous fellows raised a Prussian flag on a staff sur-
mounting Lazard's saloon on Treasure Hill. "This greatly enraged our
friend Lazard," reported the press, "who hastened to capture the offen-
sive bunting; and, placing it under his feet, tore it to shreds in less time
than we have been writing this item — meanwhile offering to fight all
the Teutons in White Pine." The editor noted that a San Franciscan
had offered a reward of $500 for the first Prussian flag captured by a
Frenchman and suggested that Lazard was certainly entitled to the
"soap."[308]

As the traveler of today walks through the abandoned cemetery a few hundred yards from Hamilton, the topsy-turvy headstones whose inscriptions can still be read reveal much about the people who made up the mining community of White Pine. The impact of foreign personalities was great. "John Young, Dumfrieshire, Scotland, died May 22, 1878, age 46;" "Peter A. Peterson, native of Sweden, erected by his friends;" "Louis C. Zadow, native of Germany," and many, many more. Grave markers also illustrate some of the personal tragedies of the camp, the toll of human life. Mary Case of Cambridge, Massachusetts, was buried in June, 1870, at the age of nineteen, and a monument was erected to her memory by an "esteemed friend," and no doubt a broken-hearted youth, Isaac Phillips. Young Harvey Adelbert Travis died in July, 1873, having lived only nine months.[309] Morris Cook, fifteen, an only son who tended his father's store in Hamilton, died suddenly in July, 1870, an event which sent "a thrill of anguish through the community." The newspaper noted, "Morris had endeared himself to all who knew him. He was a youth of correct habits and sterling principles, and gave bright promise of a noble manhood. Cut off in the Springtime of his career, he leaves a void in the circle of his family and friends that can never be filled this side of the shores of the Unknown."[310] Unlike these unfortunates, the great majority of White Piners lived to work, to speculate, and to move on to other mining camps.

LOCAL GOVERNMENT

The western mining camp is often pictured as democracy run rampant. This concept has survived because emphasis has been placed upon the leveling influences of the migration and on the miner's activity in the initial weeks when he sought shelter, located his claims, and participated in the formulation of district law. The boisterous, rollicking society, on the surface carefree, cherished democratic equality in its recreation. There was also a great deal of interest in preserving equality of economic opportunity. However, when the district survived the first two seasons, residents were forced to consider more basic institutions, like the church and school, to create permanent governmental structure, and to inculcate civic pride. If White Pine can be cited as an example, at this point participation and interest in the democratic process broke down. Political activities became the concern of the minority; only a small fraction of the mining population became registered voters. County and municipal government was supported primarily as an adjunct to the mining district, augmenting its regulations. From the first, the tax structure of local government in White Pine was inadequate, and the miners refused to pay the minimal assessments that were levied. As a result, there was always a chronic shortage of funds. Nevada counties established as a result of mining booms struggled for existence but somehow managed to survive. Then as now the patronage factor loomed large in the creation of new counties and new districts. All things considered, the democracy of the mining camp appears self-interested, politically passive except where property and patronage matters are concerned, dedicated to materialism, careless of democratic theory, ideals, and institutions, and insensitive to minority interests, like those of foreign citizens. One might characterize the situation as an immature expression of democracy, were it not that this viewpoint had been extended into the mature society of the the nation in the twentieth century.

As early as the winter of 1868, a petition was circulated among miners at Treasure Hill urging the legislature to create a new county from the eastern portions of Lander and Nye counties. The first name proposed for the new county was "Ruby," but vigorous exception was voiced because the new mining district was already being mentioned in the financial and commercial journals of New York, San Francisco, London, Paris, and St. Petersburg as "White Pine." In addition, settlers along the Humboldt River and in Ruby Valley in the northeastern part of the state were likely to procure a new county once the transcontinental railroad had brought sufficient population. That county would have a prior claim to the name of "Ruby."[311] Mass meetings held in Treasure City, Hamilton, and Shermantown endorsed the name of "White Pine" and urged the legislature to appoint a Board of County Commissioners in the act creating the county that would serve until other officials could be elected. Three men were placed in nomination: Frank Drake by Treasure City, Frank Wheeler by Hamilton, and T. M. Luther by Shermantown.[312]

Older political units, only on the rarest occasion, welcomed a reduction in their geographic boundaries or influence as new developments or population pressure on the frontier justified the creation of a new state, territory, or county. As was expected, citizens of Austin objected to the splitting of Lander County. A protesting petition, with twice as many signatures as taxpayers in the entire county, was sent to the legislature. The growing indebtedness of Lander County because of its limited tax income was the basic reason presented. The Treasure City newspaper argued, "Lander will have all the territory it ever derived any revenue from previous to the discovery of White Pine, after this new county is formed, and why they should complain or enter protest in this matter we cannot tell."[313] A bill was introduced into each house of the legislature to create White Pine County and also to make it a new judicial district. Debate continued throughout January and well into February. The state constitution prohibited the legislature from altering boundaries of the judicial districts of the state or from increasing the number of districts or judges at any time except at the expiration of a term of office or when a vacancy was created. This provision protecting the independence of the judiciary from legislative influence was used by opponents of the new county to argue that the move was unconstitutional. Advocates of the legislation insisted there was no prohibition against organizing a county and leaving it temporarily attached to Lander County for judicial purposes.[314]

The boundaries of White Pine County, as fixed by the legislature, reveal that the new political jurisdiction was literally carved out of the wilderness. The area was bounded as follows:

> All that portion of the State of Nevada lying east of a line running due north and south through the most westerly part of the house known as Shannon's Station, on the westerly slope of Diamond Mountains, in Lander county, on the road from Austin to Hamilton in said county, and south of a line running due east and west through the most northerly part of Camp Ruby, and north of the present line between the Counties of Nye and Lander, as located by Thomas J. Reed, County Surveyor of Lander county, made in 1868.[315]

Although this new county contained approximately eight thousand square miles in southeastern Nevada and included the overland road, White Piners were disappointed. They had wanted all of the eastern portion of Lander County, including Elko and a section of the Central Pacific Railroad. They argued that residents in the northeastern part of the state were still forced to travel by overland road from Ruby Valley to Austin to reach their county seat. Hamilton was fifty miles closer.

The bill also named twelve county officers only one of whom, Frank Drake, had the endorsement of local residents. Moreover, $22,500 of the indebtedness of Lander County had to be assumed by White Pine County, and 20 per cent of the new county's revenues was to be set aside in a special fund for payment of this debt. Hamilton was named the county seat. In anticipation of an increase in population, the county was allotted two state senators and five assemblymen.[316] This last provision was the only decision of the legislature approved by the majority of residents in the mining district.[317]

The legislators in Carson City had seized the opportunity in establishing the new county to reserve the jobs that were created for political friends, most of whom resided in the western part of the state. When news was received that "outsiders" had been appointed to the county offices, another petition was circulated urging the governor to assert his right of appointment by naming a second set of officials composed of local residents and let the two sets of appointees contend for the positions. The press challenged thoughtless residents requesting procedures that would bring the organic law before the courts in costly litigation thus delaying the organization of the county for at least two years. The *White Pine News* suggested, "The officers appointed by the Legislature are not to hold forever, but an election will soon be held and then we can oust them if we see fit. The evil of contesting the matter is far greater than in submitting to an imaginary wrong."[318] A patronage war was thus averted and the county government was finally organized in April, 1869.

Treasure City businessmen, sorely disappointed that Hamilton was named county seat, were particularly irked when the newspaper of the latter town suggested that Hamilton might also form a separate mining district.[319] This proposal was stillborn, but several Treasure City lawyers, smarting under the affront, did propose the removal of the county seat from Hamilton. Learning of this move through the "grapevine telegraph" the editor of The *Inland Empire* noted that eternal vigilance was the price of county seats as well as liberty. The Nevada law provided that whenever legal voters to the number of three-fifths of those participating in the last general election petitioned the Board of County Commissioners for the removal or change in location of the county seat, an election had to be held on the proposal within fifty days. The Hamilton newspaper launched an attack on Treasure City:

> Not satisfied with having the richest mines, the loftiest town, the dearest water, the coldest winds, the most enterprising Town Trustees, the bravest men, the handsomest women, the fattest babies and the finest view of the country, Treasure looks with hungry eyes on our "ewe lamb" of a county seat, raised by expensive nursing at Carson, and will doubtless make an effort to take the present chamber of justice from the green fields and suburban surroundings of Hamilton into the chilly clime where the Pogonip rages and the San Francisco speculator mourneth for his first-born.[320]

After enumerating numerous legal reasons why the transfer would be difficult, the editor advised his "townsmen who visit Treasure on business to keep a wary eye on all papers presented and not be deluded into signing a petition for an election, on the ground that it is a prayer for a new mail route or a subscription to help bury a stranger who froze to death while walking down Main Street in Treasure City."[321] The publisher of the *White Pine News* denied that the scheme for removal was being given serious consideration; he had a great deal more to say about the rival journalist:

> His slurs upon the locality are all gratuitous — nobody here has been troubling himself about the belongings, blessings nor peculiarities of Hamilton. While the mines are good, and improving, Treasure City is well fixed, and has enough business to employ her people. Whatever benefits arise from the richness of the mines will inure to the city located in the heart of the mines, its population immediately engaged in the work of development. When Treasure City gets to be so badly off as to need the county-seat as a nucleus around which to hold the settlement, Hamilton will be badly off indeed.
>
> No, we do not want your "little ewe lamb of a county-seat," Messieurs of The Empire; and it would be in better taste for you to refrain from making faces at a generous community which, though blest with all the disadvantages of climate you choose to taunt us with, is yet content to let you keep the seat of legal business two miles away from what attracted us all out here. The county-seat, intended as a convenience for the men engaged in developing the mines, ought to be in Treasure; but representatives false to the facts as they knew them, were fain to assist in your "expensive nursing" at Carson, last Winter, and this community was defrauded. But we have enough without it, and so let it rest; and keep your temper better, till you are worse hurt.[322]

The 1869 session of the Nevada legislature redistricted the state for judicial purposes to the extent the constitution allowed. White Pine County remained in the Eighth Judicial District along with Lander County. The judge, William H. Beatty, was required to divide his time between the counties, holding two alternating terms of court, and each county was to pay one-half of his $5,000 salary. An additional section added to the law revealed the preference of the legislators and indirectly offered the judge a bribe. If he elected to resign so a new district could be created in accordance with the constitution, Judge Beatty was to be assigned exclusively to White Pine County at a $7,000 salary. The governor was authorized to fill the vacancy in the Lander County judgeship that might be created.[323] Before the session was over, the northeastern corner of the state, which White Piners had hoped to include within their county, was taken away from Lander County and established as Elko County.[324]

Residents of the three largest mining camps in the White Pine District early realized that municipal government must be established to protect property and maintain order among the heterogeneous population. For a time it appeared that the legislature of 1869 would fail to act on their petitions for charters. Treasure Hill leaders resolved that if the community was denied a mayor and council with the legal right to enact ordinances, a governing committee would be appointed to maintain a police and fire department and to supervise the grading of streets.[325] In the closing weeks of the legislative session of 1869, however, charters were granted to Hamilton, Treasure City and Shermantown. The incorporation act of Treasure City, typical of all three, fixed the boundaries of the town and placed the ruling power in a five-man Board of Trustees. These trustees were to be elected annually along with a treasurer, assessor, and marshal; a person had to reside in the town three months prior to the election to hold office. The legislature appointed the first Board of Trustees in the act of incorporation.[326]

Although the county government was quickly in working order, complaints were made that the Hamilton municipal organization remained in embryo. This was explained by the fact that the functions of the county commissioners had been prescribed by state law whereas the city trustees had first to frame an entire set of ordinances for the government of the community. An able lawyer had been chosen city attorney for this purpose. The Hamilton newspaper thought the selection of a marshal to preserve the public peace was not nearly as urgent as the choice of a street commissioner who would adopt uniform grades for the town's thoroughfares, and see to proper drainage and the construction of sidewalks. "We trust that our City Fathers will see that Main Street has a sewer constructed capable of carrying away the water that has been trying, during our warm weather of late, to sluice away the entire street."[327] When action was not immediately taken, the journalist complained of the "deplorable condition" of the streets in every issue of his paper pointing out the specific location of bogs and mud holes. He also demanded that an ordinance be passed prohibiting people from depositing rubbish in the street.[328] When the city finally established a uniform grade for the major streets, businessmen and residents were urged to cooperate by putting down sidewalks, even though lumber was expensive, because "to build a good house and not have a sidewalk in front is like neglecting to put doors on a house."[329]

The date of the first municipal elections in each town was set for the first Monday in June, 1869. Within six weeks after incorporation, many Treasure City residents had expressed interest in the "loaves and fishes" to be distributed at the municipal election. Six candidates were already in the field for marshal and another six expected to announce before election day. A swarm of aspirants had also filed for the assessorship.[330] The Hamilton paper urged voters in that town to refrain from partisanship and elect the most qualified men, irrespective of their former party affiliations. Re-election of the entire Board of Trustees appointed by the legislature was recommended.[331] The newspapers of the district next launched a campaign to get residents to register by appealing to their civic duty and pride. Results were disappointing. When the registry agents closed the books in Treasure City, only 647 names had been listed. The town's editor observed, "Very little interest appears to be taken in our approaching municipal election, even by the candidates themselves, and, as a logical consequence, not half the voters of the city registered."[332] In Hamilton's two wards there were 483 citizens on the voting lists, and the editor there commented, "This may seem to outsiders something marvelous and strange for a city boasting of nearly four thousand inhabitants, and can only be accounted for upon the grounds that our people have, instead of politics, chloride on the brain."[333] Voters of Shermantown manifested as little interest as those in the other two towns and up to the final day of registration only fifty names were on the voting roster.[334]

In an attempt to stimulate voter interest in Treasure City on election eve, the *News* reported complaints that, either because of private interest or lack of proper attention, city authorities had been buying property for streets at prices not warranted by business conditions. The editor admitted that "a saint in Heaven" could not give universal satisfaction as a trustee, but warned voters that election day was the time to appraise the accusation and to "grind any axes."[335] The election was quiet and orderly primarily because saloons were closed by state law, and not until they re-opened at sundown did the usual knockdowns on the street resume. Around midnight, the canvassers announced that Thomas Coleman had been elected marshal and Stewart C. Bell, assessor. Somewhat disgruntled, the *News* noted, "It was not generally understood that there was to be a party fight, but somehow, a majority of those who are probably elected are Democrats."[336] In Hamilton, the Board of Trustees were re-elected without opposition in an election "conducted without regard to party organization or political issues."[337] The pattern was the same in Shermantown where the ninety-one voters in the election were asked to choose between two slates of officers designated the "People's Ticket" and the "City Ticket."[338]

In spite of rigid economy, all three municipal governments were hamstrung by limited financial support. By August, 1869, the City Fathers of Treasure were over $10,000 in debt and were forced to authorize the municipal attorney to take legal steps to procure a loan of $15,000 to $20,000 through the issuance of city bonds due in two years. The town's scrip was circulating at a fifteen per cent discount on the dollar.[339] Hamilton fared better. Her municipal functions had to be curtailed, but the steady growth of population and increasing business as a supply town contributed sufficient tax revenue to pay the debts. By mid-August a modest surplus of $12,000 had been built up.[340] In Shermantown the citizens were at loggerheads with the trustees over a business license tax that was thought to be exhorbitant, and mass meetings were periodically held to move toward disincorporating the municipality. The "People's Party" favoring disincorporation had been defeated at the municipal election by only a few votes. As the town's debts continued to mount, one citizen complained, "If we do not disincorporate, and if we keep on voting new taxes, we will soon be like the individual who could do no business where he was because he had no capital; and could not go away because he could not pay his debts; consequently he was of no use to himself or his creditors."[341] Some Shermantown residents who favored disincorporation would not support action in that direction because the legislature had so confused the corporate limits of the White Pine communities that the town was within the boundaries of Treasure City and likely to "fall into the jaws of another ravenous incorporation." The Hamilton newspaper, expressing sympathy for the taxpayers of Shermantown, suggested that her mounting indebtedness was the result of bad management, and proposed a close watchfulness of the men who handled the town's funds.[342]

White Pine County, born in debt, never had sufficient resources to fulfill its basic responsibilities as a governmental unit. Private schools were gradually replaced by public-supported ones when bond issues were reluctantly approved by the voters to construct meager facilities. Public institutions, such as the jail and the hospital, provided only minimal facilities. By August, 1869, scrip of White Pine County was worth only eighty-five cents on the dollar. Officials insisted, nevertheless, that incoming taxes would be sufficient to satisfy all the demands against the county with the exception of the Lander County debt.[343]

The town of Hamilton took on a more permanent appearance in the autumn of 1869 with the construction of a brick courthouse, including a court room and a jail.[344] Within a few months after its completion, a grand jury, upon investigation, reported that the building had not been constructed according to the specifications of the contract. Experts testified that the courthouse should have been erected for $31,000, while the contractor had obtained $38,000 from the county. The Hamilton newspaper commented, "As public buildings never have been erected as cheaply as private ones, we think White Pine may congratulate itself that it lost no more."[345]

Miners were exceptionally casual about paying their property taxes when due, and county and city officials constantly complained about delinquencies. At the beginning of 1870, approximately 30 per cent of the county and city taxes assessed for the previous year were outstanding, penalties had been assessed and steps were being taken in some instances to take over real estate. One newspaper editor warned:

> Unless you wish some speculator to get a perfect title and immediate possession of your valuable lots and improvements, go and settle at once. Many men are already selecting property they intend to buy at tax sale.
> You can save money by making payment immediately. The law will be strictly enforced, and you will lose your property unless you settle your tax.[346]

White Pine County also had its quota of dishonest public servants. The county treasurer, Lewis Cook, was accused by the grand jury of misconduct because of his questionable handling of county funds. Public monies were scattered in three depositories, one of which was Cook Bros. of San Francisco where cash had been placed on general deposit for the use of that California firm. Cook arranged for the return of funds in all haste while the grand jury continued its deliberations, and he escaped with a reprimand.[347] Later in the year, while his accounts were being audited, he "retired from the county," absconding with $24,000 from the treasury. A $1,000 reward did not lead to his arrest and, in the end, the loss was shared by his bondsmen and the county. About the same time, the elected mayor of Hamilton, one Harper, was discovered using public funds for personal purposes. These peculations staggered the community, for the city government was virtually bankrupt, and the county organization crippled to a marked degree.[348]

The harmony of Hamilton was further disturbed in 1870 when the Board of Trustees granted an exclusive franchise to furnish gaslight for the city streets and the homes of citizens. According to the proposed contract, no competing company could lay pipes in the city for twenty years. A maximum rate was provided in the agreement, but negotiations for the specific charge were to be conducted with the company by the city and by each householder wanting gas. Hamilton's mayor refused to sign the ordinance because he questioned its constitutionality. Other members of the Board and the gas company sought a writ of mandamus to compel him to sign.[349] The district court judge admitted that he was not clear as to the mayor's powers and he would take the case under advisement. The jurist added fuel to the flames of controversy, however, by stating from the bench that he considered the ordinance an "infamous thing" and if the law compelled him to grant the writ he would withhold it sufficiently long to enable some citizen to apply for an injunction against the grantees. The judge was disturbed by the monopoly powers granted the gas company. He was incensed at provisions that would permit the company to delay its installations, not having to commence work for nine months nor to lay two thousand feet of pipe until two years had elapsed. He suggested that if the city were to be lit by tallow candles, the trustees might, in parallel fashion, require that the contractors put the wick in the molds within nine months, and at the end of two years add the tallow.[350] *The Inland Empire* supported the mayor's position in opposition to the franchise, but the *News,* that had recently transferred its office from Treasure City to Hamilton, thought there was nothing unusual in the arrangement with the utility company. The *News* editor rebuked the judge for his conduct:

> The court unbends itself to say how it will cheat its own decision, if duty at the last compels a decision adverse to the outside views and prejudices of the court! . . . Judge Beatty means well, but circumstanced as he is, he ought to have a cooler brain. He ought to be free from the flitting fervor of partiality which leads a man to catch at so thin a device as injecting a stump speech into the remarks which announce a suspension of judgment. . . . It was not dignified. It was gratuitous. It was irrelevant, smacking of the street corner. It was wrong — the merging of the man and the partisan into the judge. . . . We were sorry to see it; and hope Judge Beatty will drop the style, and disport himself more seemly; for there is the making of a good man in him, if he will take advice as kindly as this is meant, and try to profit by it.[351]

The newspaper debate increased in its fury, the legal arguments became more technical, and the final action of the courts was delayed so long that the gas company lost interest in the project.[352]

PARTISAN POLITICS

County and municipal governments had scarcely been organized before the time arrived for the annual election of recorder for the White Pine Mining District. Early in July, 1869, ten aspirants had announced their candidacy including W. P. Tenney, the incumbent. One correspondent of the *Mining and Scientific Press* urged San Franciscans to interest themselves in the re-election of Tenney, described as a "good custodian" who had carefully kept the records. Not only had he established an efficient office in Treasure City, equipped with a good safe, but his deputies were also constantly in the "field" accommodating miners needing to file claims. A new administration, less familiar with the history of various locations, might confuse the records and create litigation without end.[353] Hamilton's *Inland Empire* did not approve of the re-election of Tenney, and demanded that the entire law code of the mining district be reconsidered at the annual meeting. The "twenty ignorant men" who had prepared the first laws had provided that no business could be transacted at a special meeting unless a majority of all miners in the district were present. Such requirements were deemed unreasonable in a district with six thousand miners. Only a majority of those present at the regular annual meeting was necessary to adopt new regulations. Hamiltonians were urged to "trudge to Treasure" and demand the appointment of a committee on by-laws to report proposed amendments to the district regulations at a future adjourned meeting. The editor also thought this legislation should not be considered at the same time as the election of the recorder. Treasure City was reasonably satisfied with the status quo; Hamilton hoped for changes in legislation and personnel. The basis of a partisan struggle was thus laid.[354]

The recorder, up for re-election, called the meeting for the early morning hour of 9 A.M. in his Treasure City office that would not hold more than a hundred men. The hour was too early for the miners of Hamilton, Shermantown, and elsewhere to arrive in time without personal inconvenience. Moreover, in the call no statement was made as to the procedure in voting or the length of time the polls would remain open. Hamilton's editor was convinced that "those who reside at Treasure Hill will vote en masse. But those of the important towns and camps, near and remote, will be but poorly represented at the single polling place in which the election will come off."[355] The Shermantown paper also insisted that at least three precincts, in addition to Treasure City, should be established to record the votes in the district. Miners scattered all over the area, twelve miles square, would be forced to leave their work and make a day's journey to cast their ballots. "Not one-third of the miners in the district are centered at that city," complained the editor, "and it is unjust and oppressive to those in distant locations to be obliged to trudge for miles over the mountains to cast their votes."[356] The Treasure City newspaper, recognizing the importance of the election, admitted in all frankness that the laws which "were possibly good enough for struggling prospectors are not suited to the permanent condition of things into which we have advanced." Requirements of boundary definition, description of claims, and declaration of bearings were known to be inadequate. However, the laws had been laid down in good faith and, in this newspaper's opinion, no radical revision that might lead to confusion and litigation was called for. When the meeting assembled, the first action should be to provide for an adjourned meeting to hear a report from a committee on laws; selection of that committee should then follow. This committee could, in quiet deliberation, determine where amendments should be inserted, redundancy erased, and ambiguity eliminated in the code. The revision could be published for the consideration of all interested miners prior to the opening of the adjourned meeting. Thus, the Treasure City editor speaking for at least a segment of his community was prepared for a considered amendment of the mining regulations, but the residents of Hamilton hoped to scrap the entire by-laws and begin over. All thinking men agreed these decisions needed to be made, free from the angry contentions of factions in a popular election.[357]

When the annual meeting convened, the first resolution provided that the recorder's office should remain permanently in Treasure City. A committee on the revision of the laws, elected to report at a later meeting, was composed of nine men, seven of whom were from Treasure City, one from Hamilton, and one from Shermantown. W. P. Tenney was then re-elected by an overwhelming majority. The Hamilton paper the following day labeled the meeting "a put up job," "a Tammany Hall trick," and "preconcerted trickery." Announcing that Hamilton and Shermantown residents were justly indignant, the editor insisted that the local crowd assembled at Treasure City demonstrated such anxiety to carry out "their predetermined programme" that "they neglected to cover their shamefully one-sided proceedings with even a veil of fairness and decency."[358] The Shermantown press agreed that the proceedings were conducted by a "ring" of company agents, officeholders, and mining speculators.[359] The editor of the *News,* however, insisted that the election resulted in the choice of a man for recorder acceptable to miners who had been working claims in the district longer than any town had existed. Treasure City, as the center of mining interest in the district was the logical location for the recorder's office. Hamiltonians were unsuccessful in electing any more members of the committee on law revision because they insisted on supporting candidates committed to removing the recorder to Hamilton. Miners of Treasure Hill were determined not to have their mining laws amended by "a committee of ranchers, wood-choppers, and vaqueros."[360] The journalist continued, "The meeting was fairly conducted in every respect, and real miners are satisfied with the result. All charges of unfairness are fabrications, having no shadow of truth."[361] Moreover, he thought it was not necessary to quarrel about the matter. Both Hamilton and Shermantown were free to set up new mining districts with laws and record offices arranged to suit their convenience.

———•————•—

The residents of White Pine County participated in a general election for the first time in 1870, and partisan politics came to the fore early in that year. The Democrats held a mass meeting of three hundred people in the courthouse at Hamilton on March 21, to organize a political party. Early in April they assembled again at the opera house to set up a county committee and to elect a representative to the state Democratic committee.[362] The *White Pine News,* Republican in sympathy, inquired "What is the Democratic Party of 1870? Has it any principles other than those of steady and unyielding opposition to everything proposed by some other party?" In true partisanship, the editor suggested:

> The policy of Jefferson and his supporters finds no place in the creed of Modern Democracy. . . . No act of patriotism, performed by the brave and loyal, is considered a virtue. The blood-stained and battle-scarred veterans of freedom receive no praise. . . . The widows and orphans of the loyal dead are opposed in their just demands for National aid, and their destitution creates no feeling of pity and receives no sympathy from the so-called Democracy.
> Can men of sober judgment and sound reason expect any good from a party that has no creed save that of hostility to the Government in which they live and of hatred toward those whom they can not control?[363]

The Republicans organized a primary election in May to select seven delegates from each township to attend a convention that would, in turn, name candidates for county office. The *News* warned, "It is time that the Republican party should realize that the enemy is in front . . . it is incumbent upon all true friends of the Republican cause to see that no sheep are lost in the changing of pasture."[364] While the editor urged Republicans to participate in the primary, he admitted in advance that the Democrats are "far better in making a large showing in the primaries than we are; but when it comes to the regular elections, especially when we have a first class ticket, they have been in the habit of badly slipping up."[365] The Republican primary was conducted without incident, but the Democrats became involved in a shooting fracas at one of the polling places of their party in Hamilton. Shots were fired by disputants, landing in the midst of a large crowd of citizens exercising their right to vote. Although members of the city police and constables from the sheriff's office were present, no arrests were made. At the next meeting of the Board of Trustees a committee was appointed to investigate the "disgraceful derringer and revolver tournament."[366]

Both parties urged all eligible men to register so they could vote in the municipal election of June and the national election of November. "Next to the yield of our mines, there is nothing so important as having good officers to manage our local affairs," suggested the *News* "and unless good citizens interest themselves in the matter, the election will go by default, and Tom, Dick and Harry will be left to run riot and rule over us."[367] In each of Hamilton's two wards, there were only thirty-eight voters registered five days prior to the deadline; the press issued frantic appeals, particularly to Republicans, not to let the election go by default.[368] In the recent Republican primary there had been 448 votes cast. Many citizens insisted the small registration was the result of the legislature's requirement of a four dollar poll tax — considered a direct violation of the Fifteenth Amendment.[369] When registration closed there were 885 names on the rolls in Hamilton. Everyone was convinced that the municipal election would be closely contested. The editor of the *News* prophesied:

> The Republicans will carry the city without a doubt, if they will take an active interest in the contest; and no time should be lost in reconciling the wavering, and in urging upon the apathetic the importance to the party of securing the prestige of victory. . . . If we triumph, we shall secure the vote of Hamilton to the Republican cause in the next general election. Up and at them![370]

On election day, citizens were urged to "vote early, but only once." The editor insisted, "You will find that in the main the Republican ticket is the best. The Democrats have some good names, and we are not going to lie for any party; but a ticket of best men must have a Republican majority."[371] When the ballots were counted, the Republicans had elected Hamilton's town officials and, according to reports, "The Chiv. Democracy and the Caucasian, or Plaza Democracy, were cursing each other in unmeasured terms," each accusing the other of bringing about defeat.[372] The municipal election in Shermantown was very quiet with only one hundred and sixty voters participating. In Treasure City the chief issue was whether municipal government should be abandoned in the interest of economy.[373]

The Inland Empire, that had ceased publication in April, 1870, was reestablished by the Democrats in October to counteract the influence of the pro-Republican *News.* G. W. Cassidy, who had worked on the *News* in the early days at Treasure Hill before Forbes had made it a

Republican organ, was placed in charge. Financial support for the revival came from L. R. Bradley, Democratic candidate for governor in the approaching general election.[374] The campaign got under way with a mass Democratic rally on Treasure Hill where the state's most talented orators appeared. Early in October the number of registered voters in Hamilton had increased to 950. With only sixty-five names added to the rolls since the books were opened for the fall election, there appeared to be little interest displayed on the part of either party. Once again the registration law passed by the Republican-controlled Nevada legislature was labeled an "infamous outrage." The four dollar poll tax was said to place a price upon the vote. The sensitive poor man unwilling to accept aid from politicians lost his suffrage and those who accepted it were under obligation to the party that paid his tax. Moreover, the foreign-born citizen was discriminated against for he had to produce papers proving he had been naturalized, submit to an "inquisitorial examination," and take oaths before getting his name on the voting lists. "Can it be," asked the editor of the *Empire*, "that the great Republican party has no faith in the foreigner?"[375]

A Republican rally at the Stone Saloon in Hamilton was described by the opposition press as "a very slim affair." Not over fifty individuals were in attendance, including a half-dozen colored citizens.[376] For two weeks prior to the election an alphabetical list of registered voters was published on the front page of each issue of the *Empire*. Members of both parties continued to rustle up unregistered citizens at the rate of twenty-five a day. The press spearheaded this movement with editorial comment like the following:

> Democrats, the time has come to organize. Fail not to know every man in your neighborhood and learn his politics. Every one must consider himself a committee of one and go to work. Our enemy is doing everything that money and work will accomplish, and we *must* work.[377]

Len Wines, the stage owner, had been nominated by the Republicans for state treasurer. With Jerry Schooling, another White Piner who was the Democratic candidate, he left Hamilton to make a joint canvass of voters of western Nevada. Charles W. Kendall, Democratic candidate for Congress, conducted a series of debates with incumbent Tom Fitch in Virginia City, Hamilton, and Shermantown. The Republican Congressman was described by the Democratic press as a "mineral land-grabber and general corruptionist." Prior to his last Hamilton appearance, he was goaded by the comment, "Surely, Fitch, thou art chief among knaves."[378] In the final desperate days prior to the election, working men were reminded that "the Republican party stands up squarely for the voluntary immigration of the Chinese," and the introduction of "cheap labor" in all the mines of the state. "The immigration as well as the importation of rat-eaters must be stopped," screamed *The Inland Empire*. "To accomplish this desirable end every laboring man should vote the Democratic ticket from top to bottom."[379] Reports throughout the state indicated that Democracy would sweep everything before it. "Let us, as one man, wrap the shroud of Coolieism around the funeral car of Radicalism and send them, as twin brothers in infamy, to a grave which knows no God or resurrection," pled the editor.[380] Such tirades continued daily until the election on November 8, when every true Democrat was advised to make certain he had a red-marked ballot with flags and clasped hands drawn on the front.

The election was orderly throughout the entire county, with very little drinking and no reported rows at the polling places. When the saloons opened after six o'clock in the evening, they were crowded with people anxious for election news. Considerable coin had been staked on local candidates and the election of the county sheriff was known to be a very close contest. Never in the history of Nevada politics had there been so many bolters from party ranks and as much scratching of the names of individual candidates. Voting had been determined more by emotional, local issues, like the suppression of Chinese labor, than by the need for reform in the Republican Party at the national level. Morning came before the victors in local races were known, and final returns from western Nevada did not arrive for twenty-four hours.[381]

For the first time in the history of the state of Nevada, the Democrats were victorious everywhere, electing governor, Congressman, lieutenant-governor, state treasurer, and judge of the supreme court. L. R. Bradley, a Virginian by birth, who resided in Lander County and was the spokesman for the eastern "cow country," won the governorship from the Republican candidate by a vote of 7,200 to 6,149. Kendall, who was the first Democrat to represent Nevada in Congress, had been born in Maine and was a graduate of Yale University. He had gone to California during the rush of 1849, edited the San Jose *Tribune* between 1855–59, practiced law in Sacramento, and served in the California legislature before going to Hamilton at the height of the boom. The Republican and Democratic forces were deadlocked in equilibrium in the state senate and assembly. Republicans did not regain their ascendency in state politics until 1878.[382] Voters and candidates in eastern Nevada, centered in White Pine County, had largely been responsible for upsetting the Republican organization in 1870. With its objective completed, *The Inland Empire* ceased publication and soon thereafter was

sold to H. C. Patrick, who removed the press to Stockton, California.[383]
Although the Republican Party, dominated by its more radical elements,
had elevated a victorious general to the Presidency and apparently held
a tight grip on the nation's politics, the numerical division between
Republicans and Democrats in the country was much smaller than generally believed. To protect the slim margin of control, Republicans never
hesitated at election time to raise pre-Civil War issues, to urge veterans
to "vote as they had shot," and to associate the Democrats with dis-
loyalty. Such campaigns were, more often than not, successful. Candidates for local office, when hard-pressed, attempted to project national
issues into state and county contests, and this sometimes also represented the margin between defeat and victory. In the Trans-Mississippi
West, party regularity was minimized in this decade, and regional interests, like the concern of the mining states and territories over mone-
tary policy or the agitation on the Pacific Coast over Oriental immigration, influenced the voter far more than issues that were the residue
of the Civil War. Nevada was Democratic between 1870–78 while the
nation remained Republican. A basic factor precipitating the shift in
1870 was the antagonism all over the West against the federal government for failure to halt Chinese migration. As elsewhere, the state's
political balance was delicate, and the Democratic majority in White
Pine was enough to make the difference. Here the Irish miners, a significant number in the population, had a predilection for the Democratic
Party. The excessively high poll tax levied by the Republican legislature
strengthened the workingman's belief that his interest lay with the opposition. These factors, coupled with the re-establishment of a Democratic newspaper to publicize the nature of the cause, and the politician's
skillful commitments concerning future patronage, combined to give the
Democrats control of the state of Nevada.

PROPERTY AND PROSPERITY

As a rule, mineral deposits were located either on the public domain
or on reservations assigned to the western Indian tribes. Since 1785
lands containing valuable minerals had been reserved from sale or preemption by the United States, so miners everywhere, in California, Colorado, western Nevada, Idaho, and Montana, were trespassers. Moreover,
there were no federal or state codes of mining law whereby property
might be protected. Custom dictated that the codes of the mining district should be recognized as legally binding. Finally political pressure
from western mining states and territories forced Congress into action

during 1866, and the local rules and customs of the miners, as prescribed in their district regulations, were recognized as federal law.
The miners were no longer trespassers and could register their claims.
They were not required to do so and since the costs, including a minimum payment per acre plus surveying and recording, averaged around
$1,000, many miners doubted that their claims were worth it. Although
the confused legal situation had been clarified just prior to the silver
discovery on Treasure Hill, most residents in the White Pine district delayed in securing a federal patent to their mining claims or on town lots.

Under both Spanish and English mining law, the vertical extensions
of the surface boundaries of a claim were the underground limits of
what the owner might mine. Californians modified the law so that the
owner of the upper part of a vein might follow it downward indefinitely
in any and all directions even if it meant tunneling under another's
property. The practice spread throughout the mining camps of the
entire West and was incorporated into the district regulations at White
Pine. Difficulties arose in every mining camp that put this principle
into practice. It was a particularly annoying practice in White Pine
because there was endless debate as to whether the silver deposits were
a geological vein or just surface outcrops.

Lawyers were legion around Treasure Hill, as elsewhere in the
Nevada mining camps, because of the confusion and the haphazard
methods in registering claims. The only reason they did not gain as
much wealth and notoriety as their associates at Virginia City was that
the stakes in litigation were not so high. The judiciary was not exceptionally able, but just as belligerent and unyielding, once a decision was
rendered, as most frontier judges. Throughout the mining kingdom it
was the practice of the unscrupulous newcomer to record a claim as
close as possible to the workings of a wealthy mine, to dig shafts into
the pay dirt, and claim the lode as his own. The established company
was blackmailed into buying off the claims of the latecomer or threatened
with a long and costly court fight. Usually these tactics gained no support, but in White Pine when such cases went to trial, public opinion
seemed likely to consider the established, older company as an aggres-
sive empire-builder. Joint-stock enterprises found the answer by obtaining a federal patent to their land.

During the height of the White Pine boom, real estate prices nat-
urally soared. Lots selling for $25 in Hamilton and Treasure City in
August and September, 1868, were worth from $600 to $1,200 by
November. Nearly every building spot along the road from Hamilton
up to and through Treasure City and along the road to Shermantown
was claimed; when selling to newcomers, the holders invariably asked

for an advance on yesterday's prices.[384] The San Francisco *Alta California* reported in February, 1869, that lots sometimes jumped hundreds of dollars in value in a single day.[385]

The streets of Treasure City were crowded from morning to evening by men rushing around with bundles of real estate documents in their hands, and "by the appearance of things a stranger would come to the conclusion that the whole State of Nevada was being transferred."[386] Henry Eno's first impression was that White Pine was "the richest silver mining district since Columbus discovered America." The district appeared to be a veritable paradise for speculators and adventurers. Some men who came had money, others had sagacity. A few were destined to make a fortune, the majority would know only failure.[387]

Several men filed claims to lots in the middle of Treasure City's Virginia Street during February, 1869, staked off their property in this thoroughfare, and hauled in lumber to erect buildings. The constable started removing the stakes and the lumber, and when the jumpers protested he drew his six-shooter to enforce the law. They had the constable arrested. The case was tried immediately before the local judge who discharged the defendant, fined the prosecutors, and forced them to pay the cost of the suit. The local press commented, "There is hardly room enough now for a team to pass through any of the streets of our town, and we hope that no one will hereafter interfere with or obstruct the few thoroughfares that are used by the public."[388]

Claim jumping became a serious problem and public confidence had to be restored. Hamilton residents asserted that it had become unsafe to leave town lots, though improved, even for a few hours, and proposed an organization "to protect all persons in their rights to property acquired in an honorable way." At its first meeting the "protective league" adopted a series of resolutions: the titles of the "original proprietors of land" were valid and possession was absolute, therefore these lands would be held until the courts proved to the contrary; the rights of squatters and jumpers who attempted to seize property would not be recognized; public peace and order were to be maintained as well as the rights of private property. The group intended to enforce these principles "by any amount of physical force that the exigencies of the occasion may require."[389]

In May, this Citizens' Protective League was called upon to sustain the resolutions it had adopted. A band of "lawless jumpers" from Montana seized some Main Street lots of absent citizens and erected a fence along the street. At an early hour in the morning about four hundred of the largest property holders in town "peacefully assembled and marched in procession, under the orders of the marshal of the organization, to the scene of action." The jumpers, impressed by this "imposing demonstration of physical force," asked for a parley and finally decided to take down their fence and tents. One observer reported, "All was done peaceably, and in good order. No violence, no bluster, no threats marked this uprising of the people, a solemn determination that spoilation and robbery should not be perpetrated with impunity appeared to pervade the whole assembly. The men of violence and rapine saw and understood this town was no good place for jumpers to try their midnight strategy."[390]

Property holders in Shermantown found that title to the entire area upon which their town had been built was legally claimed as a mining site by several men. A town meeting was called and a committee appointed to retain counsel to present the community's case at the next session of the district court. For the purpose of defraying expenses, each property holder was to be assessed by the committee according to the amount of his interest at stake in the contest. The lawyers were granted a fee of $3,500, with $2,000 of the amount contingent upon successful culmination of the suit. The News noted, "Our Shermantown neighbors are moving with great unanimity in this matter of 'striking for their homes' and will no doubt make a big fight when the proper time arrives."[391]

The attempt to organize a Citizen's Protective Union was not as successful in Treasure City as in the other two towns in the district. Here there was more interest in mining claims than in real estate for business or for homes. Not until the general prosperity was threatened by claim jumpers did the tradesmen and real estate owners sponsor a mass meeting to deal with the problem. As usual, a committee was appointed to draw up resolutions and report. Upon reconvening, the proposed resolutions were passed without a dissent by the small group attending the town meeting. Only a few were willing to pledge their support when the time came to sign the agreement, some from a lack of concern and others from a desire not to be associated with a vigilante crusade. Leaders insisted that the Protective League did not propose to use violence and did not want to be thought of as a vigilance organization. The News insisted, "It is a worthy movement, for its objective point is the ousting of trespassers, unprincipled adventurers, who lust after the fruits of honest men's toil. It will be a success, for the men are in it who have no dread of these depredators, and organize only because in union is strength." Everyone was urged to seize the opportunity to join the organization.[392]

The building boom continued into the spring of 1869. *The Inland Empire* commented:

> The sound of the carpenter's hammer is heard in all directions of this little city daily, and every morning the improvements are so clearly visible that even a close looker-out for improvements is almost astonished at what has been accomplished the day before and on the morning of his getting out of bed, if he rises late. Let the resident depart from here to-day, and return in two weeks, and he will hardly recognize the town, so rapid are improvements made.[393]

Builders never inquired about price from lumber dealers but whether or not the essential materials for construction were available.

In June, 1869, Hamilton's tax assessor announced that the town's real estate was appraised at $1,250,000. "We are not disposed to crow or say anything we cannot back up," observed *The Inland Empire,* "but we are satisfied there is not another town eight months old in the world that can show the same prosperity." Assessable property was expected to be $2 million by the time the town was a year old.[394] Many individuals made fortunes out of real estate in the mining district. One gentleman who landed in White Pine in January, 1869, with his blankets on his back and six bits in his pocket was required to pay an income tax of $3,300 six months later.[395]

————•—◄—•————

"Rich mines invariably attract the disciples of Blackstone," observed the *White Pine News*. Members of the legal fraternity had made their appearance in great numbers during the cold winter of 1868–69. The local editor commented further:

> They know that things are apt to be done in a loose way in a mining camp, and well knowing that many transactions will be backed by a lawsuit, the generous fellows would not for the world have one stand in need of their services without being able to come to the rescue.[396]

The journalist was right in complaining about the haphazard manner of recording claims. Unfortunately, laws of the mining district did not specify that when a mining claim was filed some definite landmarks be given to determine its exact position. Not one in a hundred claims filed with the recorder designated the exact *position* of the holding but merely stated that a person claimed two hundred feet along a specified ledge or lode.[397]

Clerical carelessness in the recorder's office was another source of difficulty. A case in point was the *Pocotillo Silver Mining Company* vs. *Peter Brandow* of San Francisco. Brandow, while acting as superintendent of the Pocotillo Mine, located near the eastern base of Treasure Hill, discovered that the original location notice of the mine claimed a vein running south when it was intended by the discoverer to have claimed to the north. He immediately erected a claim on the valuable ore deposit and had it recorded in the name of other parties who then conveyed it to him. He resigned as superintendent of the Pocotillo and proceeded to San Francisco where he was about to sell his claim when an injunction was issued to forbid the sale on the ground that his title was not clear. The District Court later ordered the injunction dissolved. A new case was then instituted to determine whether the company's title or Brandow's was valid. Both parties to the litigation expended more on these court cases than the mine would have earned, and in the end were forced to compromise their differences. This case, and others, served as a warning to locators to be careful in wording their notices, and to be explicit in all descriptions.[398] So many conflicts in title developed that claimants, weary of paying legal fees, adopted the policy of consolidation. The Treasure City newspaper thought, "It will be severe on the lawyers, but consolidation is cheaper in most cases than litigation. Let the work go on."[399] The lawyers, nevertheless, enjoyed a good business. By summer, 1869, there were forty-four names enrolled on the list of attorneys in White Pine County, "comprising, doubtless, a greater array of legal ability than any other bar in the State."[400] One careful observer, who had served as a judge in California, reported that a small army of lawyers was on hand when the district court was in session. Trials usually lasted from ten to twelve days. Only a few of the lawyers had a thriving business and most of them "depended more upon perjury and subornation of perjury than upon principles of law or precedents" to win their cases. Experts on mining did a thriving business as witnesses, reaping larger profits than the lawyers.[401]

Miners were usually so impatient to reap the rewards of their claims that few bothered to get a land patent from the federal government. Commenting on the Congressional law of 1866 providing that ground containing precious metal could pass in fee simple to the occupant, the *News,* as a spokesman for White Piners, remarked, "We always doubted the need of the law, and doubt now the advantages. It made some business for officials, who have to be paid by the miners." The miner wanted nothing from the government but the right to work his lead of precious ore. Before 1866, it was argued, he had the right, on the basis of mining district laws, to work, use, let, and sell his claim and no one ever dared dispute his right to a properly registered claim. Under the "absolute title" law the way was opened for the outsider who arrived late in the district to file a claim for fee simple title and force the miner into court to defend his rights. The miner was given an outside chance by the law to check the sharpers,

> . . . if he can ever get through the hurdle of official requirements and legal proceedings, to enjoy — unmolested — the identical rights which before were guaranteed him without any such trouble. It is true, as the framers of the bill will here interpose, he has, after going through the hands of a surveyor, a lawyer, the Surveyor-General, and the Receiver at the Land Office, and the Department of the Interior at Washington — if he doesn't die of old age or over-drafting — he has a tangible title, which is just as good to the ground after it is worked out, as it was while the quartz lasted. *That* is something.[402]

The Department of the Interior required twelve documents for certification of title.

During the August-September term of the federal district court, the judge disposed of sixty-five cases. Fifty-nine untried cases were left on the calendar for a second term that was convened within a week after the close of the first.[403] Two cases, involving the Aurora claims, each lasted nine days, and important decisions were handed down. The court ruled that a single location consisted of only two hundred feet, and that two days' labor had to be performed for *each* location, and the same recorded, to hold the claim one year. The vast majority of claims, originally made by partners, had only two days' work performed to hold the whole, no matter what the footage, and the owners had recorded but two days' work. The judge refused to recognize the custom of the district or the fact that not more than ten or twelve owners out of over a thousand had recorded more than two days' work. He was not impressed by the lawyers' offers to present testimony from the founders of the district as to their intent, but insisted that the laws they had written must prevail. Virtually all mines held in San Francisco were subject to relocation, and many miners feared the loss of their property.[404] The Supreme Court was requested to review the case in February, 1870, but had not handed down its decision five months later, much to the annoyance and bitterness of many property holders.[405] In a replevin suit of *South Aurora* vs. *Autumn* to recover bullion taken by the latter company from ground in dispute between them, the court refused to give judgment for the plaintiff. The judge insisted that the defendant, the Autumn Company, having obtained possession of the property under a title equally good as that of the other party could not be held to have unlawfully obtained possession of the ore, until its title was judicially pronounced invalid. The fact that a suit was pending to determine the title of the mining ground from which the bullion was taken did not affect the rights of either party to possession of the bullion when that possession was obtained without fraud. The *News* hailed this decision as "proper and just" because the South Aurora Company had been the aggressor. The editor suggested that the companies with old, established title were accustomed to pouncing upon newcomers who happen to strike a rich claim nearby with a restraining injunction and then try to trace their original lode into the new discovery and claim the property. The editor commented further:

> Such a course retards the development of the mines and drives capital away. The more just and proper course would be, for those who think the ground occupied by another party belongs to them, to first prove it by tracing in a proper manner; and then the title will be the more easily settled by the courts and without doing an injustice to anyone.[406]

When the federal court recessed for the winter, there was an exodus of lawyers from White Pine. One newspaper commented, "A large number of our legal gentlemen have gone to California to spend the holidays."[407]

————— • ■ • —————

Miners arriving at Treasure Hill were immediately faced with the problems of housing and subsistence. There was no hotel in the district during the winter of 1868–69, but Wakefield and Wheeler's store rented single bunk beds with a hay mattress — crickets thrown in — for a dollar a night. This general store was temporary housing for most newcomers and the man who timed his arrival so as to secure a bunk was fortunate. The less favored slept on the floor of the store, in saloons, restaurants, tents, or in the open air. A meal of slaughterhouse steak and coffee was worth a dollar. Board in Hamilton cost from twelve to fourteen dollars a week.[408] The prevailing wage in Virginia City, where many of these men had worked, was four dollars a day, and in some Nevada camps daily wages were less. Weekly expenses for housing and food for the newcomers amounted to between 80 and 90 per cent of the wages they had been accustomed to making elsewhere.

Men fortunate enough to locate a claim often lived nearby in a cave. Those who became wage earners would have preferred to build a small cabin of lumber in one of the three towns, but lumber supplies were limited and the price prohibitive. Neighboring sawmills provided wood at $175 a thousand feet but by the time it was hauled to Treasure City the price was $200. At times when the supply was short, prices went up as high as $400 a thousand. Only the tradesmen erecting mercantile establishments could afford such prices. Desperate for shelter, groups of a half dozen men joined forces to put up shanties of cedar posts. The outsides of these dwellings were usually chinked with stones and mud and roofed with cedar boughs and earth. The few who erected buildings of planks often lined them with cotton cloth to exclude the wind, and used either a shingle or tin roof.[409]

Fuel was another essential along with housing and food. Because of the cold weather, fires were kept burning in the miners' cabins every day and night except in midsummer. Firewood cost $3 to $5 dollars a jackass load. During the spring of 1869, the hills for eight miles north of Hamilton were alive with wood choppers. All the woodlands surrounding the district were being stripped and the wood corded for winter use.[410] Several Chinese spent the summer gathering cordwood and stacking it on Pogonip Flat in anticipation of cold weather. The pile increased daily, and a newsman noted, "John will make his pile next winter."[411]

Prices were notoriously high for tools, household equipment, and food in all mining camps because of their isolation and the uncertain and expensive transportation by freight wagons. The circumstances whereby freighting operations led to inflation in the White Pine District have already been outlined. Scarcity kept prices high until April. Flour sold for $16 to $17 a hundred pounds, and Nevada potatoes were worth twelve and one-half cents each. Essentials like coffee brought seventy-five cents a pound and sugar was three and one-half pounds for $1. Candles needed to light the inside of darkened shelters and mines were thirty-five cents each. There was no credit at the bars where liquor was sold for twenty-five cents a drink. Beef was more reasonable than most items, selling for thirty-five to fifty cents a pound. Other meats were prohibitive in price; chickens sold for $5.[412]

Business boomed without regard to the high prices prevailing. *The Inland Empire* reported:

> Yesterday was the liveliest day we have ever seen in Hamilton. From an early hour in the morning the streets were filled with people wherever they could find standing room. All kinds of teams, from one horse carts to sixteen mule teams, were crowded together. Horse auctions, Cheap John auctions, and the public sales of all kinds of property seemed to be in its glory, and buyers were abundant and good prices paid. We visited a number of blacksmith shops and found miners waiting for the sharpening of their tools, and were informed that the work in their shops was now double what it was a month since, and increasing every day, although many new shops have gone into operation and nearly all the leading mines have their own shops.[413]

One newcomer noted that everything except a horse was salable in Hamilton. Keeping animals was too expensive. Hay sold for twenty-five cents a pound and barley for thirty-three cents. One man, forced to keep horses for four days in Hamilton prior to disposing of them, discovered his expenditures had been $84.[414]

Because of the exceptionally high cost of living, minimum wages at White Pine were a dollar a day more than they had been at the Comstock even during good times. Wages paid in the mines were $5 a day in coin, and any man not working a claim of his own had no difficulty finding employment on those of others. Carpenters were better paid, earning on an average of $6 to $7 a day, and in a few instances $10. In an attempt to cut labor costs, some mining companies constructed boarding houses and fed their workers. At mills where the men received no board, the chief engineer received $8.00 a day; engine drivers, $6.00; rock handlers, $6.00; and pan men, $5.50. As wages rose during the working season of 1869 above those of the previous year, prices also went higher.[415] Hamilton's newspaper told the following story to illustrate that even the Chinese were "getting cute."

> Yesterday a restaurant keeper in this City, wishing to hire a Chinaman to wash dishes, made his request known to a head Chinaman. In a short time an almond-eyed son of the sun made his appearance and was asked by the proprietor what he would expect. John replied he wanted $25 for one week, and was told it was too much, but he would give him $20 a week. John's reply was: "No, me no work for twenty dolla; you can no hire Chinaman for twenty dolla; you hire plenty Melican man for twenty dolla." It would seem the Chinese are getting high toned, and are willing to work for such wages as are paid Americans. What next.[416]

Luxuries were almost non-existent and, when available, prohibitive in price. On noting cherries in the market during May, one journalist remarked, "As we did not have money enough to buy half of one we did not make any inquiry in regard to the price. It was some enjoyment to look at them and think back to the days of boyhood when we could eat them and they had no taste of silver. To eat them here would be like eating first class White Pine ore."[417] Until mid-April, eggs hauled in from Salt Lake City cost $2 a dozen, but upon the arrival of a wagon load from Paradise Valley they were being hawked in the street at $1 a dozen.[418] By the end of June the price was cut in half again when Young and Maupin, storage and wholesale merchants, received one thousand-dozen eggs, along with several tons of potatoes, at their stone store building on Main Street in Hamilton.[419]

Supplies of food were plentiful throughout the summer. All kinds of fruit — cherries, apples, peaches, apricots, grapes, oranges and straw-berries — could be obtained in Hamilton. In Shermantown, ox-trains appeared weekly from the Mormon settlements loaded with potatoes, sold by the sack at five cents a pound. Eggs now were abundant in this community, as elsewhere.[420] Prices were also more reasonable. Flour could be obtained for $8 a hundred pounds, beef and mutton fifteen to twenty cents a pound, bacon thirty cents a pound, sugar three pounds for a dollar, and beans, ten cents a pound. Water still was sold at the high price of twelve and a half cents a bucket; wood was $6 a cord. Because of the shortage of building materials, rents continued high, running anywhere from $40 to $400 for a one- or two-bedroom shelter. A drifter noted that there was "plenty of good air, but of a rather light quality," for nothing.

———•———•———

By spring of 1869, prospectors had located claims all over the three peaks of Treasure Hill, in Chloride, Bromide, and Pogonip Flats be-tween the peaks, and were moving down the hillside.[421] The official claim records of the White Pine Mining District have not been located, but one local historian states that the mining recorder was forced to employ three assistants during 1869 and 1870 to keep the statement of claims up to date. Within two years an estimated thirteen thousand claims had been filed.[422] The total number of valid titles never reached this figure at any given time because many individuals filed a half-dozen or more claims, one at a time, abandoning each as it proved valueless. In spite of the feverish activity, no new discoveries were made com-parable to those of 1867–68. Many prospectors, disappointed in Treasure Hill, moved back to White Pine Mountain, where the first discoveries in the district had been made in 1865.[423]

Silver was taken from the earliest claims consistently, but estimates of the amount and value of bullion produced vary widely. Three sources of statistics were available: reports of superintendents of the White Pine mills, accounts of express companies and forwarding agents, and the rolls of the county tax assessor. One conservative estimate of the total production during 1868 was $500,000, but this figure must have been very low because a single mill in Shermantown reported processing $235,144 worth of silver in the last two and one-half months of the year.[424] Another published compilation of production between May, 1868, and the end of the year totalled $968,813.[425] It is quite likely that as much as $1,500,000 worth of bullion was produced during this first year because quantities of silver were processed that were not recorded.

In the early spring, agents for Wells, Fargo in Hamilton, Treasure City, and Shermantown adopted the policy of releasing monthly reports on bullion shipments.[426] Comparisons of their figures with mill reports indicate that many silver bricks were being fabricated that were not entrusted to the express companies for delivery. Two Shermantown mills announced that their silver production for the first three months of the year was $297,023, although one of them had only operated for a single month. This was a remarkable accomplishment, for this ore had all been excavated during the winter. Local journalists insisted that the White Pine district was shipping out more bullion than any mining district on the Pacific slope except Virginia City.[427]

Many successful miners of White Pine carried their fortunes to other localities for investment. The lucky discoverers of the Eberhardt, Hidden Treasure, and South Aurora all purchased real estate in Oakland, California, and elsewhere along the east San Francisco Bay region. One of the discoverers of the district, T. J. Murphy, paid $50,000 for a half interest in the Broadway Block in downtown Oakland. Others had invested as far south in California as San Bernardino. Local residents thought a portion of these fortunes should be expended in advancing the "rugged mountain communities" where the riches were obtained. "We expect, however," wrote the local editor, "that Treasure Hill and Eastern Nevada will furnish much more treasure to the world than it will receive."[428]

When Henry Eno reached Hamilton in 1869, he reported that many handsome fortunes had been made the previous season and many had also been lost. The largest fortunes had been made by men who already had money, but there were some poor men, like the discoverers of the Eberhardt, who were thought to be millionaires. Governor Henry G. Blasdel of Nevada had realized $200,000, and Leland Stanford and his brother had used their position and capital to make much more. The California Bank of San Francisco owned several mines and was operating a twenty stamp mill.[429]

Available mills could not handle a tithe of the ore in March, 1869, and until adequate facilities were constructed residents in the greatly over-populated district were certain to devote their time and energies to speculation.[430] The editor of *The Inland Empire* told what was happening:

> We have good mines and many of them, but it is a rare thing they are sold. The most of the mines thrown into the market are those which are located for speculation, and are either worthless or else they are locations made on ground belonging to other companies. Those buying mines should remember that a good mine

in White Pine is not to be purchased for a thousand or two
dollars, for it would take but a ton or two to pay that amount.
The truth is, there have been a vast amount of locations made,
and the parties making them have sent them to San Francisco
and incorporated them there, and have an office on Montgomery
Street, and have flooded the whole country with a lot of trash
which is not worth the paper it is printed on. . . . We do not know
of six good mines in this district that have been incorporated.[431]

The first White Pine companies, eleven in number, were incorporated
in California during December, 1868. In January, this number more
than doubled when twenty-six were registered; February showed an
increase of over one hundred per cent; then eighty companies were incorporated in March, nearly as many as in the previous three months.
In all, there were 169 companies, representing a capital of $246,884,000
divided into 2,330,061 shares. The Consolidated Eberhardt and the
Eberhardt Mill and Mining companies, legitimate enterprises, had the
largest capital, $10 million and $12 million respectively. The San Francisco *Bulletin* published a list of all companies and the report was
reprinted in other city newspapers and by the *Mining and Scientific
Press*.[432] This mining journal was taken to task by White Pine newspapers for quoting figures of authorized capital of incorporated companies as though it was money actually invested. At least two hundred
millions of the nominal capital in these White Pine companies did not
represent legitimate, paying, mining claims. Everyone knew that the
district was not producing sufficient bullion to pay the interest on such
an investment in watered stock, and speculators were thereby undermining the reputation of the district.[433]

Speculation was not confined to the San Francisco exchange, for
at the beginning of 1869, a Stock Board had been organized in Treasure
City to facilitate the exchange of shares in the White Pine mining companies that had taken over the major claims. The *News* protested the
action, pointing out that the primary purpose of stockbrokers was to
force as many sales as possible to collect commissions. For example,
any shareholder of a mining company could employ the Stock Board to
sell a portion of his interest at a sacrifice to depreciate the value of
the rest, so he could buy the additional shares necessary to secure control of the property at a low price. Brokers were reminded not to "kill
the bird that laid their golden egg."[434] The warnings of the press did
not check the activities of the Board. By March, seats, limited to fifty,
were commanding a premium of from two to three hundred dollars.
Business was so flourishing that sessions had to be held daily rather

than semi-weekly, as in the winter months, and more elegant accommodations were being sought on Treasure City's Main Street.[435]

Although capitalists were warned to be cautious about mining properties in the district, they were continuously urged to invest in badly needed milling facilities. As elsewhere in the West, mill construction at White Pine had lagged behind ore production during the first season of mining activity because time was required to disassemble a structure in a declining district and transport it by wagon to the new discovery. Because of damage to machinery in transit or too hasty construction, breakdowns of the stamp mills were frequent. Foundries and machine shops in California had difficulty filling orders for all the parts that had to be installed or replaced. Ores of the Treasure Hill mines were, for the most part, simple in character. Most of the silver-bearing material was chloride of silver that could be readily worked by means of amalgamation, known locally as the "Washoe process." Since the chloride ores required no roasting, smelters at first seemed unessential. However, crushing mills had to be constructed to pulverize the ore for the amalgamation process. The average cost of milling chloride ore was from $12.34 to $14.60 a ton, with $4 to $5 of this amount being paid for the haul from the mines to the mill.[436] As more carbonates were uncovered, smelters were also in demand.

In February, 1869, there were only four quartz mills in operation in the district: the Oasis, Moore's, Treasure, and White Pine. The Monte Cristo mill was not in repair. Moreover, the splendid Newark mill, twenty-five miles away, was idle because of financial difficulties. Reports circulated that the Big Smoky mill of Kingston and the Old Dominion mill of Hot Creek had been disassembled and were rolling in wagons toward Hamilton to be reconstructed there.[437] Of the available mills, the Oasis owned by the Eberhardt Company and located in Shermantown was the best. With its two batteries of five stamps each, the mill capacity was about twelve tons a day. Eight large settling pans were used in the amalgamating department and adjoining this was the retort room containing four furnaces capable of retorting two thousand pounds of amalgam at one blast. Near the Oasis, the company had purchased the Drake and Applegarth mill running eight stamps and capable of processing ten tons a day. Both these mills, like most in the district, were run on the dry crushing principle.[438]

At the height of the 1869 season there were nearly two hundred mines within an area four miles square, producing profitable ore, and it was estimated that an additional dozen mills would find full-time employment.[439]

Early in the summer of 1869, ten mills with one hundred and twenty stamps were steadily at work on White Pine ore. Milling capacities of the district were expected to double during the course of the next three months. The Consolidated Chloride Flat Company was erecting a second mill of twenty stamps. The Aurora Mining Company was considering the construction of a large mill of thirty stamps. The fifteen-stamp mill of the Metropolitan Mining Company was almost ready to open. In addition, the Pogonip & Othello Company had recently uncovered such a large amount of ore that the proprietors were considering building their own mill. When these mills in process of construction were completed, there would be a total of 185 stamps.[440]

By autumn, geologists working with the King survey reported fifteen mills at White Pine running anywhere from nine to thirty stamps, with capacities from five to fifty tons a day.[441] The following list of mill facilities, available or in construction, prepared at approximately the same time, confirmed the claims of the geological survey.

Drake and Applegarth	8 stamps
Manhattan	24
Metropolitan	15
Moyles'	5
Monte Cristo	10
Henderson	5
Kohler (or Staples)	8
Nevada (Dunn & McCone)	10
Oasis (Eberhardt Company)	16
Treasure Hill Mill (Big Smoky)	20
Social and Steptoe (Egan Canyon)	20
Treasure Hill (Murphy's)	5
White Pine (Miller's)	10
Shaw and Felton's	5
Newark	20
Total number of stamps	181
Under Construction:	
Stanford	30 stamps
Dayton	20
Mineral City (Robinson District)	10
Tregloan's	10
Vernon	10
Total	80 [442]

A few months later the Nevada State Mineralogist reported an increase in the number of mills to seventeen, with a total capacity of 214

stamps.[443] At one time twenty-three mills were running simultaneously.[444] In addition to the mills located in Hamilton, three other milling
centers of importance had developed. Capitalizing on her bountiful
water supply, Shermantown had provided facilities for five quartz mills
and four furnaces by the end of 1869. The camp of Swansea, near
Shermantown, had also been selected as the site of two mills and a
furnace during the summer. Six wagons arrived in the town bringing a
portion of a third mill all the way from Alpine County, California.[445]
Eberhardt had the largest mill of all, the Stanford, with thirty stamps.
In 1870, the Eberhardt and Aurora Mining Company was erecting a
sixty-stamp mill there.[446]

In the mountain range immediately west of Treasure City, large
quantities of low grade silver ore which was refractory had been discovered. It contained enough carbonate of lead so that heating in a
furnace was necessary for the silver to be reclaimed. The ore from this
Base Metal Range, yielding from $60 to $100 a ton, could be processed
at a profit only with adequately equipped furnaces.[447] Soon a smelting
business grew up separate from milling operations in the community.
Rothchild's Smelting Works at Hamilton operated three furnaces. The
works were housed in a brick building with smoke stacks forty-five feet
high.[448] In the district there were eight additional smelters of various
capacities.[449] White Pine did not suffer from a lack of knowledge about
milling and smelting techniques. The ways and means of dealing with
chloride and carbonate ores, unlike sulphurets and more complicated
geological combinations, was both simple and well-known.

After all the disparagement of the district during the spring of
1869, the mining press of San Francisco reported a renewed interest in
White Pine during the summer, partially because of the discoveries in
the Base Metal Range. Money was more available in the Bay City and
the mills had produced convincing proof of the presence of precious
metal in the shape of silver bricks. The *Mining and Scientific Press*
proudly announced, "A great majority of the producing properties are
already owned in this city as well as a large minority of the non-pro-
ducing and 'untamed feline articles.' "[450] Mills were reportedly overrun
with ore and had contracted with mine owners for all they could handle
in the next three months. Teamsters refused to visit out-of-the-way mines
because they had all they could haul from the claims easily accessible
to the mills. Sizable shipments of bullion continued throughout the
summer months. The Hamilton paper warned, "Look out, Comstock,
your laurels are in danger. For August we will ship at least one and
a half millions."[451]

Like all frontiersmen, residents of White Pine were primarily concerned over their material well-being. Owners of mining claims undoubtedly were asking themselves whether their footage was worth working and if so, whether the legal costs to hold on to it were justified. Merchants with expensive and sizable supplies speculated on whether or not high prices could be maintained so they would recoup investments. Laborers hoped that prices would come down before their exceptionally high wages. Investors and mill operators disagreed on the amount of bullion that would continue to flow from the mines of Treasure Hill. In short, all were carefully considering the unanswered question concerning the length of time prosperity could be expected to continue.

CHAPTER THREE

THE DECLINE

In early spring of 1869, San Francisco newspapers suggested that the populace of White Pine was overdoing its promotional work. For example, the *Alta California* noted:

> Real estate has been run up to such a point in this place that it is now of very low sale, and there is a lull in the market. Four or five thousand dollars for an ungraded hillside lot, 25x100 or less in size, on a street which is ungraded and almost impassable, is too much for the majority of people to stand, and holders will have to modify their views or run the risk of finding themselves permanent holders of a very unproductive property.[1]

More thoughtful members of the community finally recognized that real estate speculation had gone too far. Although land speculators were ever present on the Western frontier, their activities had a tendency to check the development of a community. All land within a mile or more of Hamilton had been surveyed and applications to many sections had been filed under the preemption law of the United States. These locations on the public domain had obviously been made for speculative purposes and such action was illegal. Large tracts of land were lying idle while new arrivals in the district were being driven away by high prices. Some argued that if holders of these tracts had been willing to sell every other lot at a reasonable price to a settler who would build on it, the actual value of the remaining lots would have more than doubled.[2]

By the close of the summer of 1869, the district was overstocked with supplies and prices began to topple. The cost of "grub and whiskey" was among the first to be affected. Restaurants reduced the price of meals from one dollar to fifty cents and, "in view of the stringency of the money market," some saloons reduced liquor to twelve and a half cents a drink.[3] Local ranchers began to provide more meat than could be consumed. A "beef war" broke out intermittently among competing butchers and the populace took advantage of the situation.[4]

The one exception to the declining price structure was lumber. After a depressed market through the summer months, this trade started to boom in the fall of 1869 when residents began to build stronger shelters against approaching winter. Lumber was scarce, particularly in Treasure City, and orders sent to the yards in Hamilton could not be filled. No weatherboard was available. Prices again rose to $300 a thousand and the lumber merchants were doing a good business at that price.[5] The cost of building was almost prohibitive, and completely out of line with the rest of the economy.

THE STRUGGLE OF LABOR

The independent prospector of the type known in California survived in White Pine for only a few months. During the first season those who had not uncovered an ore body sufficient to pay someone else to do the work of extracting it, found that they had to go to work immediately to meet day-to-day expenses. Prices in all Nevada mining camps were so high and fluctuating that the miner had enough only for necessities. If there were profits they went to the freighters, and to saloon and shopkeepers who were careful to appraise the market and not overstock.

Miners on Treasure Hill organized a union in April, 1869.[6] At that time men employed in the mines still received $5 a day for ten hours work. For surface work on unopened mines, workers were customarily paid the prevailing daily wage for eight and sometimes for six hours' labor, when the weather would not permit a longer day. The first disagreement between management and labor developed when the foreman of the Consolidated Chloride Mine suspended work in an underground compartment and placed workmen normally assigned there to work on the surface sorting ore. When required to put in ten hours for surface work, they left the mine. Great excitement developed among the laborers over hours of work on hearing of this incident, and rumors began circulating about a general reduction in wages. The foreman received a note suggesting that he leave the country. The Treasure Hill newspaper warned the Miners' Union not to be influenced by the "thoughtless and and reckless counsels" of agitators. "Threats will accomplish nothing, and violence will not be tolerated for a moment in this community."

The editor elaborated upon his views:

> Some irresponsible persons in this community are endeavoring
> to foment discord, and arouse an antagonism between capital
> and labor. . . . Without capital there can be no employment —
> no labor performed; and unless labor can be obtained by capital
> for reasonable remuneration, capital will be compelled to seek
> other fields for investment and labor for employment. The interest
> of the laborer is almost identical with that of the capitalist — both
> being dependent for employment on the success of the enterprise
> at hand. If, then, labor seeks to extort from capital prices or con-
> ditions which would be ruinous to it, the investment ceases to be
> profitable, and must and will be withdrawn until other and more
> profitable arrangements can be made.[7]

Conservative counsel prevailed in the union and its membership grew
steadily. By summer, over one thousand mine workers out of twelve
hundred employed had joined the organization.[8]

On Saturday evening July 11, 1869, the Eberhardt Company posted
a notice that on the following Monday the wages of all miners would
be reduced from $5 to $4 a day. This surprise announcement was received with consternation throughout the district. The *News* cried out
against the management of the Eberhardt Company noting, "It is a great
pity that six men, owning what has been the representative mine of the
richest District in the world, cannot afford to pay the rate of wages,
which is paid by other mines in the District, many of these having never
yet realized a dollar." The management's suggestion that a large quantity of rock worked the previous month had yielded only $15 a ton
was dismissed as a "frivolous pretext." At best, the mine was not managed as well as it might be if, in the absence of the foreman, ore sorters
at the dump couldn't reject barren rock to avoid furnishing the mills with
limestone. The editor had a theory that the reduction of wages was a
"put up job" to "bear" White Pine shares in the stock market. The
Eberhardt was not in the market, for the fortunate owners had no stock
to sell, but spreading the idea abroad that the Eberhardt had reduced
workmen's wages because the rock would not pay would surely drive
down the stocks of the Hidden Treasure, Pocotillo, Chloride Flat, and
other companies. In his annoyance at the owners of the Eberhardt, the
journalist came to the defense of labor by saying, "To put down the price
of labor on hard-working miners, just now, while all other owners are
willing to pay five dollars, is a permanent wrong."[9] The *Evening Telegram* of Shermantown expressed respect for the "good sense and generous disposition" of the owners of the Eberhardt and elected not to
attribute their actions to "sordid and selfish motives." If they were

involved in a stock-jobbing operation, it would only expose their ignorance and stupidity. Most important of all, the editor insisted he was

> . . . not yet prepared to believe that the ore of the Eberhardt has so deteriorated that an honest working of the ore will not pay for working it at the present rate of wages, or that it has involved the necessity of reducing laborers' wages.[10]

The Inland Empire described the action of the Eberhardt Company as the "first grand manoeuvre" of a scheme to discourage Eastern capitalists from investing in White Pine mines, and ultimately to starve impecunious owners into selling for a trifle. In addition, the Eberhardt owners hoped to reduce wages to the rate prevailing in Virginia City, where living costs were cheaper. Was this the beginning of a "ring" like that which dominated western Nevada? The editor noticed that "this shameful blow at the reputation of our mines" was well-timed to coincide, in dramatic precision, with the arrival of Chicago capitalists representing five hundred million dollars.[11]

Many White Pine men had worked long enough in California and the Washoe mines to experience both the power and the problems of organized labor. At Virginia City, a "protective association," formed in 1863, blossomed into a Miners' Union the following year, with the primary goal of maintaining a uniform wage of $4 a day. Within the year the strength of this group was broken by the mine superintendents, not by reducing wages but by refusing employment to union members. Two years later the workers seized an opportune time to re-activate their union and were able to restore and maintain the standard wage despite a gradual reduction in living costs.

Emboldened by the success of Comstock miners, the White Pine Miners' Union instructed its members not to work for reduced wages. Some threatened to march, seven hundred and forty strong, to the mines, and "hoist" out anyone who agreed to work for less than $5. However, most employees at the Eberhardt reported on Monday morning agreeing to work for $4. Others arrived on the scene and agreed to replace any strikers at the $4 wage, and a few were willing to work for $3. Desiring to retain the old force and not wanting to precipitate any violence, the management decided to let the mine lie idle until an agreement could be made with the union. The initial attitude of labor leaders was also conciliatory. The press noted, "Although there are some hotheads in the Miners' Union who talk of violence on the street, the Union, as an organization appears to be acting with commendable coolness and deliberation."[12]

The Eberhardt Company had not acted independently but in agreement with other mine owners of the district. This leading mine had been responsible, in the eyes of all owners, for keeping wages high, and once it reduced the pay of workers others could follow suit. The Aurora South and Aurora Consolidated immediately reduced wages by one dollar. Rumors were afloat that the Consolidated Chloride Flat Company would soon follow suit as well as the Treasure Hill Milling and Mining Company that owned the Summit, Nevada, and North Iceberg mines. Faced with this united front of capital, the Miners' Union held a special meeting to consider its action. The Treasure Hill newspaper warned, "What counsels prevail or what is the spirit in the Miners' Union, we have no means of knowing, but the talk on the street of numbers and outside parties having axes to grind is very injudicious, and not at all calculated to enlist the sympathies of law-abiding citizens."[13]

On the third day following the action of the Eberhardt Company, the intention of the union was clear. Organizing a procession in Treasure City, the members marched to the Silver Wave Mine where three miners found working were forced to the surface. Reassured that these men were not employed on daily wages, the group moved on to the Hidden Treasure Mine where the superintendent told them his company could pay $5 and he had no intention of paying less. After giving three cheers, the party walked to Aurora Consolidated where the only men working were using giant powder and were therefore entitled to $5 according to the management. So the workers moved from mine to mine cheering when the foreman appeared cooperative; when superintendents elsewhere revealed an intention of going to $4 a day as soon as all would agree, they stalked away sullen and bitter. The newspapers thought "the whole proceeding was marked by a degree of quiet and good order highly creditable to the Miners' Union."[14] Two men had been selected for spokesmen at the outset and all others remained silent during the interviews with the management of the mines.

The *White Pine News* still insisted that the action of the Eberhardt Company was unkind when it could better afford to pay the old wage than any other. The editor asserted, however, that a reduction in the ruling rate of wages was now inevitable. Miners would do well to accept it, he thought, and try to negotiate a compromise settlement at $4.50 a day. "We think an appointment for a conference, with a view to compromise and resumed work, will receive the attention of the mine owners. At all events, miners should seize the advantage to be gained by asking it."[15] The *Evening Telegram* did not agree. This newspaper had shifted its position toward the mine owners, for the editor now thought the object of the San Francisco companies behind this scheme was to create a situation whereby they could introduce Chinese laborers in large numbers to reduce labor costs. The publisher commented, "We have witnessed no event of late that has caused so much disgust among the people as this first maneuver of designing capitalists for personal benefits."[16] Shermantown was disturbed by persistent rumors that the Miners' Union was planning a march into that community to stop the mills from running unless wages of five dollars were restored.[17] However, quiet reigned on Treasure Hill. The Hamilton editor reported:

> During a residence of nearly six months in this district, we do not remember to have seen so few people astir on the hill as was observable yesterday. The reverse of this might reasonably be expected, with so many men out of employment, but it would appear that the miners have taken to the woods, their cabins and elsewhere, excepting on the streets, to await results. We hear it now rumored, by outsiders principally, that the miners will accede before long to the terms of the employers; but this we consider very doubtful. The present state of affairs cannot last long, however, and the sooner the terms are agreed upon and operations resumed the better it will be for all concerned.[18]

After two weeks of enforced peace, new rumors began to circulate that the Miners' Union in Treasure City had resolved to parade again and compel men who were not getting five dollars a day to stop work. Should this be done, the owners had resolved to shut down the few remaining mines still operating. *The Inland Empire* now joined the *News* in urging the workers to accept the inevitable; "capital is king, and the republic of labor cannot by force tear one jewel from its golden crown." The editor had an explanation for the miners.

> The world is full of laborers. The mine owners have the whole world to draw from to supply the places of those men who seem determined to rule their employers, but who are but preparing the way for their own ruin. We hope the rumor is exaggerated. We hope some definite understanding may be arrived at between the mine owners and the miners. This is the critical time in the life of White Pine.[19]

Only the *Evening Telegram* continued to back the workers' demands insisting that the "wholly unnecessary action" of the Eberhardt Company had brought a "deep curse to the country." The newspaper did not deny management the right to run its business as it saw fit, but insisted that "a decent regard for the interests of the country and the character of our mines would have dictated a different policy." However pliant the Eberhardters had been to the pressure of speculators, to the great injury of White Pine, labor was warned that the remedy was not in violence or unlawful proceedings.[20]

The striking miners staged a second march on Treasure Hill to demonstrate the force of the union on July 27. All the principal mines were visited and four-dollar men everywhere were forced to quit work. Teamsters engaged in carrying ore to the mills were prevented from loading their wagons. The Stanford mill in Eberhardt City was shut down. There was talk of visiting the mills in Shermantown for the same purpose, but nightfall was approaching and the project was abandoned. *The Inland Empire* reported that miners and mine-owners appeared equally obstinate. Miners striking on the Comstock Lode in Storey County had an advantage over the owners because a large milling interest was damaged by shutting off the supply of ore. Moreover, the mines there would fill with water if left idle. In White Pine the workers had no similar advantages. In any protracted game of "freeze-out" there could only be one result.[21] The editor of the *White Pine News,* who had accompanied the miners on part of their march, was incensed to learn that he was regarded as a "spy," in view of the fact that he was a union member of long standing. He reported that at some mines workers had refused to join the union and had defied its authority; that the marchers refused to accept the word of the superintendent of the Aurora mine that all men working for him were making five dollars and demanded an affidavit; that slanderous accusations were made against other managers; and that there were suggestions about "hanging one or two" of the workers. He concluded, "We refrain from commenting upon this affair, further than saying that there were many good but misguided men in the procession yesterday. The former parade brought out 346 men; yesterday there were 106."[22] Meanwhile, 540 men were wanted for work in the mines of Treasure Hill. The support of the *Evening Telegram* for the union now collapsed. "While we have denounced the course of the mine owners for . . . bringing on this unfortunate state of things, we can not countenance disorder and force as a means of retribution . . . no association has the right or is expected to redress injuries to the entire community."[23] Once again the *Empire* issued a plea to the men to go back to work. Unemployment for thirty days meant a loss of $120 for each worker even at the reduced wage scale. For a thousand workers, the editor argued, this meant a loss of one hundred twenty thousand dollars never to be recovered. Even if labor won the strike in another two months, the losses from unemployment could not be recovered in a year's time. Employers meanwhile were suffering comparatively little from the miners' refusal to labor. Fearful about rumors of planned violence, the editor warned that nothing would be accomplished. His plea reached a hysterical pitch:

You are not the only men who are poor, and who are afflicted by earthly troubles. All round you are thousands of men struggling with poverty, trying to make for themselves a little competence, and for their families a home. The blows you strike at order and the security of the community, rebound on these; and thus you not only wrong yourselves, but through you, mothers and wives and little children are made to suffer. . . . Your chief grievance is against rich men. Do not forget that, before any course you can adopt that will bring to such men disaster, all the community besides will suffer ten fold more than they will. . . .

As a sincere friend we appeal to you to return to your labor; to live with the closest economy; to save your money. By so doing, in three months, if you then deem your pay insufficient, you can go to one of the outside districts, Kern, Sacramento, Patterson, Reveille, or Tem Piute and operate for yourself. Even at four dollars per day, in a year you can each save more money than you could in five years in the Atlantic States; more, indeed, than half the farmers in the Mississippi Valley can command after ten years of toil.[24]

The arguments of journalists inspired several union leaders to organize a "counter league" of men who were willing to work for $4 and to protect those who would work for the reduced amount. Within twenty-four hours, 150 Treasure Hill miners had signed up. The *Empire* predicted, "This, of course, will lead to a collision if another demonstration is ever made, and most probably result in a riot, bloodshed, and general disaster to the country." The newspaper once again counseled moderation on all sides to avert violence and urged a speedy conference to bring about an understanding and compromise between miners and mine-owners.[25] The Miners' Union met in Treasure City and named a conference committee of five to meet the owners and representatives of the mines at the old White Pine Theater in the basement of the International Hotel. Knowing that the unity of the union had already been broken, owners held firm to their decision to pay no more than $4. Members of the League next approached business men in Treasure City for their candid viewpoint concerning proper action to be taken, in hopes of enlisting these tradesmen's support against the mine owners. According to the press, everyone tried to convince the miners that their demonstrations were wrong. Apparently the leaders finally accepted the fact that they were beaten. Many bolters from the union, eager to go to work, had signed up to labor for $4 and several mines were ready to reopen with a full crew. The sheriff promised that no worker would be molested.[26] On the last day of July, White Pine newspapers announced that the "labor trouble" was settled and henceforth $4 would be the prevailing daily wage in White Pine.[27]

Peter Leonard, president of the Five-Dollar League, protested the newspapers' statement that affairs were settled and that there would be no further opposition to men who chose to work for $4. The *News* insisted that its statement was correct. Leonard was reminded that a majority of the workers thought the strike had "played out — that they would like bigger pay, but they begin to doubt whether proper management of a mine will justify it; and anyhow, steady work at four dollars, coin, is a much more sensible way of putting in time than parading the streets, spending in nonsense what means they have, and setting back the development of the country by interfering with those who need to work to support themselves and their families."[28] The determination of the miners to go back to work had infused new life into all branches of business and "satisfaction and contentment reigned supreme."[29]

At three o'clock in the morning of August 3, twenty-odd disgruntled miners, including the officers of the Five-Dollar League, raided three of the Treasure Hill mines and, using clubs and pistols, drove off the men working the night shift. All the attacking party wore masks. At the Hidden Treasure shaft, shots and stones were poured down among the workmen. When they came to the surface, some of the men were assaulted with clubs; others who ran were shot at. Joseph Gerrans, a Cornish miner, fell into an open pit, fifteen feet deep, while fleeing the scene and broke his thigh. Similar incidents also occurred at the Nevada and Summit mines. When news of the outbreak became known around dawn, there were threats of lynching. More moderate counsels prevailed after the Treasure City marshal arrested Leonard and other leaders of the union. The sheriff arrived from Hamilton and took six leaders down the hill to the county jail. Several men who had played an active part in earlier demonstrations of the union caught the stage for Elko. Meanwhile, the records of the Miners' Union were seized, including its roll of membership.[30] The *News* admitted that no one had expected this "overt act of outrage." "Now the law *must* be enforced, and we are glad that a clue has been struck which promises to clearly convict several of the leading spirits of the mob movement."[31] The *Empire* also denounced mob action and cried out, "Their coming was unheralded and their conduct the most reckless and outrageous ever perpetrated in a civilized community."[32] The *Evening Telegram* reported that popular indignation was boiling, that citizens were incensed at the continued outrages.[33] Fear was expressed that incendiarism might be resorted to, so a large force of private watchmen had been employed to be on the lookout.

A citizens' meeting was called the day following the outbreak. Those assembled in Broker's Hall, Treasure City, adopted measures to "insure the peace," protect the citizenry, and secure a just punishment of the "cowardly who attack the defenseless miners."[34] An executive committee of twenty-five went to work and additional arrests were made, bringing the total incarcerated to eleven. The sheriff finally wired officers at Treasure City, who were busily rounding up raiders, not to send him any more prisoners because the county jail had reached capacity.[35] The community remained tense. Each night many citizens "listened to hear an alarm of fire or the reports of pistols telling that a row had commenced somewhere," but they listened in vain. No raiders ventured to show themselves and quiet again prevailed.[36]

Realizing that public support had been alienated, the Miners' Union assembled in Treasure City and was disbanded. Before final action was taken, the presiding officer defended the original conception of the Miners' Union as "the most noble on earth."

> Its purposes were charitable — to afford assistance to their brethren when overtaken by accident or disease, or when death should lay its rude hand upon them, and cut them off in life's prime, leaving to the cold charities of the world a dependent family. Then came that noble relief to clothe and educate the orphan, to support the widow, and weave together those tender offerings which entwined themselves around the grave of a fallen brother and co-worker.

But in White Pine the object of the union had been perverted; censure and distrust was the result. Although the recent "raid" had been made without the knowledge or approbation of the union, its untenable position made it advisable to disband for the present and perhaps organize anew at a future time.[37]

Trial of the captured raiders was held in the New Melodeon Theater in Hamilton the day following the breakup of the union. After listening to evidence all day, the jury retired around nine in the evening. As the hours passed and no decision was forthcoming, friends of the accused went to sleep on the sawdust floor. Just before midnight the foreman announced that the jury was split six to six and agreement appeared impossible. The judge informed him that a verdict could be guilty either of inciting a riot or for the lesser offense of unlawful assembly. The jury retired a second time and within twenty minutes returned a verdict of guilty of unlawful assembly. Counsel for the defense asked for immediate sentence by the court, whereupon each man was fined $25 and court costs.[38]

Piqued by this decision, the *News* attacked the union. "The organization served to bring reproach upon as many good men as were connected with it, by reason of the damnable villainy done in its name." The editor reported further:

> We have been asked to make favorable mention of those who

took pains to disband the organization. It is too late. This may be said for them: that, at the late hour at which the movement for dissolution was made, they could not have done a better thing. It was late — our advice had been scouted, and threats had been made; and the bulk of those men hold us no good will now. So far, the case is even — for they owe us none. . . . We are glad things are no worse; but we have no congratulations to exchange on the subject.[39]

The press and the capitalists hailed the new prosperity destined to come to White Pine with the reduction of costs at labor's expense. Mines were being worked with a full labor crew, reduced wages at the mills would reduce the cost of ore reduction from $35 to $20 a ton, capitalists were certain to become more interested in investing, and the increasing volume of work at mine and mill would provide jobs for all. The *News* that had, a month earlier, harshly criticized the owners of the Eberhardt mine for reducing wages now reversed its position stating that it was quite right for this company to lead off in lowering wages and prices.

It will unquestionably redound to the good of the entire community, augmenting our laboring population at least one-third, and the amount of money expended in the district by outside parties in a still larger ratio. In fact, we know of nothing that could have tended more to promote our prosperity, as a people, than this adjustment of miner's wages in accordance with the prices now generally prevailing here, and we think those most forward in the movement are entitled to credit for having accomplished so much for the public welfare.[40]

Thus was the lot of the nineteenth century laboring man. Wages and working conditions were largely determined by capital and management. When wages were cut, there usually followed a division in the ranks of labor about what course of action to follow. In contrast, if there was disagreement among employers about what course to pursue in dealing with workers, the divergence of opinion was never made public. Whenever any segment of labor began to demonstrate or use force to maintain wages or obtain other objectives, community support almost always began to dissipate. The average citizen backed labor organization as long as it confined itself to fraternal, social and humanitarian objectives, but a struggle for economic advantage often resulted in alienation of the public. Knowing of the division of opinion within the ranks of labor and society, the capitalist usually sponsored counter-organization, like the $4 League at White Pine, and this precipitated violence between two segments of the workers. Intimidation and conflict inevitably led to a reaction against labor's objectives as well as its methods.

At White Pine, in western Nevada, and elsewhere in the United States, the role of the newspaper was crucial in any labor-capital struggle. Opinion was sharply divided as to whether or not the editors represented the interests of business or of the community. Perhaps the two were inseparable in these decades. In this mining camp, the newspapers constantly urged the mine workers to be cautious and conservative. Although the owners were severely criticized for the initial cutting of wages, it was concern over the economy of the entire community rather than the interests of the workingman that prompted the press to question the independence of capital. This indirect support of labor by the press did not last long. Gradually the editors counseled the acceptance of the inevitable. When labor did not oblige, the press became hostile, even inflaming public opinion against the union. Capitulation by the union became inevitable. The Comstock miners fully understood the nature of their difficulties after the initial clash with the mine owners in 1865, and exercised care to avoid conflict when they later reorganized. At White Pine the miners did not get a second opportunity.

THE PROBLEM OF WATER AND THE PERIL OF FIRE

Water and wood were always essential for a successful mining camp. White Pine apparently never lacked wood for a fuel supply either for homes or mills, nor was its procurement a major operation as at Virginia City. The water supply was quite a different matter. While both Virginia City and Esmeralda suffered from too much water that flooded their mines, White Pine had a dearth of liquid. Water was desperately needed for drinking purposes, to run the mills, and in case of fire, to stop the conflagration. Fire was the greatest threat to man and property in any mining camp. The continual search for water and the efforts to prevent fire in White Pine were inseparable, and the prosperity, even the life, of the community largely depended on the success or failure of the campaign.

In the winter of 1868–69, water from a few natural springs and dug wells was peddled on foot in the town of Hamilton at a rate of twenty-five cents a bucket. At Shermantown the rate was the same. On top of Treasure Hill there was not a drop of water except that made from snow melted over fires. One journalist noted, "This snow-water produces irritation of the bowels, and is considered dangerous if used too freely. Whiskey is recommended in its place. In justice to the people of the district we must say that they generally show a disposition to manfully conquer their prejudices, and, restraining the temptation to indulge in water, swallow whiskey with as much grace as if they liked it from the start."[41]

Water was plentiful on the slopes of Mokomoke Hill, across Applegarth Canyon, east of Treasure Hill. Plans were soon underway to pump the supply in pipes to Hamilton and, by means of hoisting stations, to raise it twelve hundred feet higher to Treasure City. In November, 1868, Colonel D. W. Von Schmidt, construction engineer of San Francisco, was surveying the ground in anticipation of building a water system. His plans involved construction of two engines to lift the water to the very top of Mokomoke Hill where a reservoir would be built for its reception. After obtaining a franchise for the water along Illipah creek, on the eastern slopes of the Mokomoke Ridge, Von Schmidt left for San Francisco in search of capital. The water works became a pet project of Hamiltonians.[42]

At Shermantown, T. M. Luther, an assayer for the Eberhardt Company, organized the Silver Springs Water Company to run a tunnel to tap several underground streams of water that had been found while well-digging.[43] After tunneling 900 feet, the men struck a stream, thirty-one feet underground, that produced fourteen inches of water—enough to furnish an adequate supply to a dozen ten-stamp mills.[44] By May, 1869, the project was nearing completion, and investors had offered the company a sum larger than its entire expenditure for a half interest in the enterprise.[45] This water resource gave Shermantown an advantage over other camps in the mining district and the local newspaper did not hesitate to admit it:

> Shermantown can boast of the coolest and best drinking water in the State. Whilst all the water consumed in Treasure City is carried there by wagons from Rock Spring and Hamilton, which becomes warm by the time it reaches its destination, necessarily causing the use of ice, having to pay good figures for both the water and the ice, [sic.] We have an abundance of water in our wells at a temperature that will make the teeth ache to drink it.[46]

Water was not only desperately needed for household consumption and the operation of the mills but also for fire-fighting purposes. The Treasure City editor noted, "The dangers of fire in our new town are apparent to the most casual observer." During the winter months of 1868–69 buildings on the hilltop were all constructed of combustible material, there was no water except that collected in buckets, and all knew that they were defenseless against fire. Although a municipal government had not been organized, citizens urged the establishment of a fire company to act as inspectors and watchmen. Property owners on Main Street, a thoroughfare densely crowded with wooden structures, considered the raising of a subscription to pay for this service. The fire watchers could also serve as policemen.[47]

The greatest danger of fire came from two sources: upset oil lamps and carelessly installed stovepipes. An alarm was raised in Treasure City during February when the oil from an overturned lamp ignited in the Mammoth Saloon in the midst of a performance by a melodeon troupe. The fire was extinguished, but the heart of the business section narrowly avoided destruction.[48] In May, the building occupied by the First National Bank of Nevada in Treasure City caught fire from sparks flying out of an inadequate stovepipe. This incident fortunately occurred at high noon and employees and passersby were able to put out the blaze. Main Street was literally thronged with people as soon as the alarm was sounded; the newspaper editor remarked, "A larger crowd we have never before seen on the streets since our stay in White Pine."[49]

The fire hazard was particularly acute during the windy season when stovepipes were occasionally blown from roofs while the residents were inside, often asleep, and oblivious to the danger. One Hamilton correspondent volunteered a solution, "Perhaps the most important suggestion we can make to the people of this city is that everyone having a stovepipe in his house should have all the joints riveted together, or at least all that portion extending from the ceiling to the top. Every little gale of wind that comes along blows down from two to ten stovepipes, endangering thousands of dollars worth of property, when the expenditure of a few dollars would prevent all danger."[50]

In the spring, the *White Pine News* launched a campaign to get the Board of Trustees of Treasure City to pass an ordinance for "inspection and close watchfulness of our city against fire." This proposed ordinance would regulate the construction of stovepipes and charge each resident with the responsibility of keeping a cask of water in readiness to prevent a conflagration. Failure to comply with the law should be made a criminal offense. The marshal and the town police should check on fire hazards, "a duty of far greater importance than the arrest of trifling and noisy brawlers, or of drunken loafers sleeping in the mud or on the sidewalks."[51] Volunteer fire-fighting outfits had been organized in both Treasure City and Hamilton during April. A Fire Committee was created in the latter town to solicit funds for essential apparatus. The town's newspaper suggested, "Let everyone who has any property in town remember that the money they pay in is simply to buy tools for others to use in defense of their property. The 'laddies' are willing to battle with the fiery element if they are furnished with tools, all they ask for their labor and dangers they pass through is the usual pay of a fireman — abuse and no thanks."[52]

Constant vigilance was necessary because of continued carelessness in allowing rubbish to accumulate and in disposing of hot ashes containing live coals. A special policeman patrolling Treasure City early one Sunday morning discovered a fire starting behind one of the town's restaurants. After the departure of the Saturday night crowd, the cook had thrown hot ashes from the stove on the ground behind the building; rubbish had blown across the coals and ignited.[53] A pile of sacks and baskets on the roof of another restaurant kitchen ignited on one occasion. A group of Indian squaws sitting in the back yard at the time gave the alarm and the Fire Department arrived on the scene to save the building. The newspaper again called the attention of the City Engineer to rubbish piled high on the roofs of several buildings on Main Street and to back yards that were full of it.[54]

Incendiaries were also at work periodically. One resident of Hamilton discovered that a large quantity of hay had been collected and placed next to the rear wall of his house and set on fire. The flames were discovered and quickly extinguished.

> We are informed by our police officers that scarcely a night has passed during the week that similar attempts have not been made. What the object of all this is we are unable to explain; but certain it is there are fiendish scoundrels in our midst who seek to destroy the city. Our citizens should be on the alert, and in the event of any of these scoundrels being detected in their nefarious designs, they should be treated to an overdose of cold lead.[55]

On Sunday night, July 12, 1869, a fire broke out in the northeastern portion of Treasure City destroying four or five wooden and canvas buildings, at an estimated loss of between $3,000 and $4,000. The Hamilton newspaper reported:

> The fire was plainly visible from this city, and hundreds of people were out to witness it. Many were about starting up the mountain to the rescue when it was observed that the fire was fast dying out. The very idea of a fire up at Treasure, where there is not water enough to make a decent cocktail, carries with it some terrible fear, and we very much doubt if an alarm of fire in our own midst would have created so much excitement as that on the Hill Sunday night.[56]

San Francisco capital finally provided a permanent water supply through an incorporated enterprise known as the White Pine Water Company. Illipah Springs, about three miles east of Hamilton, produced 2,500,000 gallons of water a day in the dry season and 5 million gallons a day in the early summer; steam pumps were erected at these springs that could handle 1,500,000 gallons every twenty-four hours.[57] Plans called for water to be pumped from the spring, through twelve-inch pipe made of riveted boiler plate, to a reservoir 470 feet above. Here a second pumping station was to raise it another 465 feet to a more spacious reservoir. Then the water was to be carried through a tunnel, 307 feet long, passing under the crest of Mokomoke Ridge, and opening on the other side upon Applegarth Canyon.[58]

By mid-April, Colonel Charles S. Bulkley, who represented Von Schmidt, had completed laying the heavy pipe on the eastern slope of Mokomoke Range, and had dug 133 feet of the tunnel. Boilers for the engines had been shipped from San Francisco. The populace eagerly awaited completion of the line for "with a good supply of water along the line of the projected mill sites, we shall certainly soon be enabled to reduce the valuable ores which we have now ready for crushing."[59] Date of completion for the project was set at July 15. A local editor noted in early June, "So quietly has this great work moved along that we have scarcely heard of it, and few in the community have any conception of its magnitude."[60] During July the twelve-inch pipe was being laid down the slopes of Mokomoke Ridge and across Applegarth Canyon toward Hamilton at the rate of a thousand feet a day. Long lines of ditches were dug ahead of the pipe layers and approximately one hundred and fifty men were employed by the company as machinists, masons, or laborers.[61]

Although the completion deadline was not met, the White Pine Water Company strained every nerve to get water into Hamilton and Treasure City at the earliest possible moment. The water flowed into Hamilton on August 14, and the company had erected a large tank near the outlet of the main pipe where water carts could be supplied.[62] These carts were used for a few weeks until pipe lines could be laid throughout the town. John Murphy, "as jolly an Irishman as ever twirled a shillelah or swigged a pot of poteen," had the honor of being the first man in Hamilton into whose house "the water of Illipah found its way in an unbroken channel." The newspaper reported that Murphy was so proud and joyful that he promised to drink nothing but water in the future for if "he had deviated from this rule slightly heretofore it was only because the Hamilton well water was not healthy." The editor observed:

> Think of it; a few months ago this was but a wilderness; now a gentleman touches a faucet in his residence, and an engine miles away answers to his touch, and over a rugged mountain pours out for him a draught of pure water. The world is moving surely, and Murphy says it goes splendidly by water power.[63]

Once the water supply was available in Hamilton, the city fathers were urged to make arrangements with the company to have public hydrants and fireplugs installed. Sprinklers attached to the hydrants could be used to keep down the dust of summertime on the streets. The municipal government was reminded that the water company was a business proposition, that over $400,000 had been expended in supplying the district with water, and that daily operational expense of the works was heavy. The city should expect to pay a fair price for the water. At the same time, the company would find Hamilton an advantageous site to construct its foundry and machine shops. The largest number of mills, all of which would be extensive users, would be located here.[64]

The White Pine Water Company requested a franchise from the city of Hamilton to lay three-inch pipes throughout the municipality. An ordinance providing for protection of these pipelines was also thought desirable. In return for this privilege, the company would make available its entire water supply system without any charge to the city in case of an outbreak of fire. The company also promised to keep full at all times any cistern or reservoir that the city elected to construct for fire protection. When the Board of Trustees of Hamilton decided that the exigencies of the town demanded a larger water pipe the company would install it, provided the municipality agreed to pay the additional expense of replacing the three-inch pipe. A committee of three councilmen, appointed to review the water company's proposition, refused to accept it and demanded instead the immediate installation of six-inch pipes. The mayor of the town, P. C. Hyman, submitted a minority report insisting that the larger pipe could not be manufactured and procured from San Francisco in less than ninety days and the city would, in the meanwhile, not have adequate water for business purposes or fire protection. At this juncture, C. P. Head, superintendent of the water company, announced that operations in Hamilton would summarily be suspended, and the business of the company hereafter conducted in Treasure City. Citizens of Hamilton were stunned by the news. The editor of *The Inland Empire* commented, "We believe that no Board of City Trustees should ask a private corporation to be at the entire expense of improvements which are not of a private nature, but wholly for the public good."[65] The water company claimed it had been keeping the mills supplied with water for a week at the loss of a hundred dollars a day. Without additional income from Hamilton residents or the municipality, the company was a losing proposition. A public meeting was called to reconsider the Trustees' refusal to allow the company to lay its pipes in the streets of Hamilton. A "large and earnest audience" assembled in the town's Opera House, and "the moral effect of public opinion when properly demonstrated" proved so irresistible that one of the trustees, who had voted against the company's proposal, agreed to change his vote. He was "deeply impressed with the unanimity of sentiment prevailing" and interpreted the proceedings as instruction from his constituents.[66] Pipes were rapidly laid throughout the town and *The Inland Empire* proposed a toast: "The days of the water carts which were filled from quagmires are over. Illipah is Water-King in Hamilton from this time onward. Great is Alla, but greater is Illipah, and Colonel Head is his prophet."[67]

Meanwhile, the pipes continued to crawl up the side of Treasure Hill. The company was making all haste to get to Treasure City with its main pipe, seven inches in diameter, to convince millmen that enough water could be delivered so they could erect works immediately adjoining the mines. The Hidden Treasure company had announced plans to build a mill at its mine site if the water could be relied upon, so that ores could be milled during the winter months without resorting to transportation through the snow.[68] The water supply coming from the Hamilton cut-off was raised 400 feet up Treasure Hill by its own pressure and forced into an inverted syphon. Pumping stations were to raise it an additional 682 feet into Treasure City. By the third week of September, the pipeline had reached the Hidden Treasure Mine, but further construction was delayed awaiting the arrival of additional pipe from San Francisco.[69] On October 9, Colonel Head finally turned on the water in Treasure City. The press again had nothing but praise for the promoters:

> It has been a tremendous labor, magnificently accomplished. It required a lavish outlay of money, and an enlightened persistent labor. It has been freely bestowed, and last night Col. Head and Col. Bulkley were, as they had a right to be, two of the happiest men in White Pine. Treasure people, justly appreciating the blessings bestowed upon them, proposed to jubilate a little. Where are our citizens? Can they not join the Treasurites in the feasting and the dancing, and wind up by all getting intoxicated on Illipah?[70]

The company constructed two reservoirs in Treasure City, one with a capacity of 30,000 gallons, and until three-inch tubing could be laid throughout the town, a water wagon was used to make house deliveries.[71] His work completed, Colonel Head departed by stage for San Francisco.[72]

Construction had scarcely been completed when the water company had to face problems of maintenance and protection of its property. Some "dastardly, malicious person" took a pick and systematically destroyed every joint in the pipe leading from Hamilton along the hillside north of town toward the Dayton mill. Replacement was not only costly but the resumption of service was also delayed.[73] The system had just been restored to working order when Hamilton's first fire occurred and residents were well aware of the importance of the water supply. The fire originated in a bakery that was entirely consumed. When the flames were first discovered around midnight, the nearby city jail had to be torn down by the "Hooks" to prevent the fire spreading and all prisoners had to be released. Several other structures were given similar treatment. Firefighters attached hoses to all the hydrants of business houses along Main Street to keep water on the blaze, and displayed "herculean efforts" in confining the fire to the quarter of its origin. A crowd of two thousand watched the event; damage was estimated at $5,000.[74]

When winter came, many residents of Treasure City melted snow instead of buying water, so the water market slumped. At the same time, the water froze and burst the pipe leading into town. The company announced that it cost fourteen dollars a day to keep the engine going to raise water to the Treasure City level and unless the town's residents guaranteed the use of twenty dollars worth of water daily, damages would not be repaired.[75] In March, 1870, the White Pine Water Company announced its decision to shut off the entire water supply in Hamilton and Treasure City. The expense of running machinery to pump the water exceeded the proceeds from sale by $4,000 a month. This action was a severe blow to the milling interests of White Pine whose operations depended upon the water. Residents of Hamilton and Treasure City resorted once again to carrying water in carts from the wells and springs in the area. Promoters of the water company had expected to supply a community of forty to fifty thousand people when they invested $400,000 in the works; when the population did not grow during 1870, the owners, recognizing their error and financial failure, hoped to recoup some of their losses by finding a buyer. The entire works was offered to the highest bidder.[76]

The Inland Empire immediately denounced the water company as a monopoly. A correspondent suggested that a "mining ring backed by the heartless money power," after securing control of several mines, now proposed to use a monopoly of the water supply to depreciate the value of other properties in the district that they expected to buy for a song. Bullion production was to be stopped by shutting off the mills, thus depreciating the value of all mining property. Many doubted that the company had been losing $4,000 a month. Hamiltonians were paying three-fourths of one cent a gallon and residents of Treasure City two cents a gallon for water taken from company tanks. The *Empire* editors estimated the cost of running the works at $2,800 a month. A group of businessmen had offered to lease the works for $2,000 a month but could get no response from the company. Moreover, if the owners were prepared to sell the works for $40,000, as the *News* reported, purchasers were available to hand over the money in forty hours to stop this "bearish" activity against the district. The editor accused the *White Pine News* of being "the organ of this company."[77]

The *News* fought back, observing, "Rings are bad enough, as they go, now-a-days; but the people who talk in circles are worse. Public interests are better off in the hands of sharpers than in the councils of jabbering numbskulls." The newspaper admitted that affairs of the company had not been managed well. Dull times had affected collection from water users. Pipes in the yards of Hamilton dwellings were being tapped by individuals who did not pay for the privilege. No meters had been put on the pipes to check the volume used. Labor had not been adequately trained to keep the machinery in operation and the few qualified workers had been putting in expensive overtime. Nevertheless, during January, 1870, the company had to draw $6,000 from San Francisco to make up the difference between receipts and expenditures for the month. Receipts for February fell $1,433 short of balancing the books.[78] In concluding the controversy, the newspaper suggested:

> It is a common custom, too generally patronized by the Press, to harangue continually about monopolies, . . . We would like to see corporations compelled to consult the public interest in advance of their own; but the world has not yet come to that.[79]

The *Empire* responded by stating that there was sufficient water in the company tank in Treasure City to last for a week and if water was not flowing by then the people would depend upon water carts to haul liquids up the mountain side as of old.[80]

———•—◆—•———

Whereas volunteer fire companies had been enthusiastically supported by the populace throughout 1869, they were severely criticized for their inertia in the spring of 1870. On one occasion Hamilton was awakened at midnight by the whistle of the Big Treasure Mill and the town's fire bell. Mill workmen had discovered a nearby blacksmith shop and an adjoining stable in flames. While workmen at the mill fought the conflagration, the firemen failed to appear. The newspaper reported that "a few, and very few, of our firemen went to the engine-house." Four members of the Hook and Ladder Company drew the truck out of the house in readiness, but only a single member of the Hose Company and two in charge of the Hyman engine arrived. As the streets were a sheet of ice, the few firemen present were afraid to take the trucks down the thoroughfares. Many firemen and citizens ran direct to the fire without going after the apparatus. Some did distinguish themselves by tearing down a wall of the stable thereby checking the spread of the flames. Nevertheless, the blacksmith's shop, together with about $8,000 worth of hardwood and tools and a new wagon worth $1,200 were destroyed. Total loss was estimated at $12,000. *The Inland Empire's* editor caustically remarked:

> This shop contained nearly all the hardwood in the market, and is a serious loss to the community as well as to the proprietors. It was unfortunate that the Hooks were not on the ground, as all the material of the stable could have been saved from the flames. . . . Had the Engine and Hose Companies been there a sufficient amount of water could have been taken from the gulch to have put the fire out in a few minutes. . . . We do hope that the next time a fire occurs our Fire Department will at least show a disposition to render whatever assistance they can.[81]

The White Pine Water Works were sold by the sheriff on April 4, 1870, to satisfy a claim against the company by the Bank of California. The purchaser, William B. Bourne of San Francisco, largest stockholder in the Original Hidden Treasure Mine, paid $43,000 in gold coin for the system. The property was turned over to F. A. Benjamin, agent for the Bourne interests and superintendent of the Stanford mill.[82] Immediately the cry was "Let us have water!" Benjamin announced that the engines would start pumping water twenty-four hours a day if there was sufficient demand from residents of Hamilton to cover expenses. If not, water would be pumped for only twelve hours a day to supply the district's mills and Treasure City. The *News* continued to lament the feeling in Hamilton that the corporation was a "soulless monopoly." City authorities were urged to confer with Benjamin to obtain an adequate supply of water for fire protection, and the town's property owners to think of the general welfare of the community.[83] Subscriptions finally were sufficient to turn the water into pipes leading to Hamilton homes, but there was no municipal water supply.

The reorganized company was primarily interested in the water supply to keep mills in operation, and soon resolved to lease the Treasure City section of the system to a sub-contractor who would be fully responsible for its maintenance and repair. Residents of the mountain town resented the profits that would go to a middleman and protested the company's plan to dispose of this branch of the service. To forestall this development, Treasure City citizens subscribed to one thousand gallons of water a day and the water company agreed to deliver it at their residences and places of business at four cents a gallon, twice the earlier price. The

new superintendent commenced repairing the line and made arrangements for carts and teams that would distribute water from the town's
tank to the subscribers.[84] The company had driven a hard bargain, but
Hamilton and Treasure City again had water.

A group of public-spirited citizens attempted to raise funds by subscription to provide Hamilton with an adequate municipal water supply
to fight fires during the summer of 1870. The fund-raising drive proved
a complete failure. The city fathers were urged to come to the rescue
and provide tax funds for a monthly payment to the water company. It
was considered suicidal folly to allow the town to stand drying in the
summer sun without an adequate water supply; it was unfair that only
fifty-two subscribers who had contracted for water deliveries to their
property should carry the financial burden, and the water company
could not afford to supply water at a loss. The press demanded, "Make
property pay for its own protection, and do not tax a public-spirited
man with the protection of a mean man's house."[85] Shortly after an
agreement had been made between the town of Hamilton and the White
Pine Water Company for the essential water, a group of taxpayers circulated a petition asking the Board of Trustees to rescind its action. In
indignation the company manager thereupon shut off the water supply,
but was prevailed upon to restore it until a public meeting could be
called. Up to this time, the water company had kept a horse and man
ready, on the first tap of the fire bell, to ride to the station and release
a powerful pressure of water. The *White Pine News* thought the demand of
the company for compensation from the municipal corporation was just:

> Whether a fire comes or not, we should be prepared for it, and
> we can not expect such protection for nothing. If the water should
> be shut off, our Fire Department would be utterly helpless in the
> event of a conflagration. We do not mean to say that in such a
> case the company would refuse water to meet the emergency, but
> time would be lost, and before a stream could be got under way,
> it would be too late to save the city. It is a penny-wise and pound-
> foolish policy to let so dangerous a risk hang over us.[86]

The matter was finally compromised and the town of Hamilton obtained
essential water.

During 1872 a number of small fires were reported in Hamilton
with losses from $200 to $5,000. The most serious fire was at the new
court house, but the Liberty Hose Company and the Hamilton Hook and
Ladder Company, still volunteer organizations, quickly extinguished
the blaze. On the morning of June 27, 1873, fire again broke out in the
center of town and was soon beyond control. All but two business
houses were entirely destroyed; losses were estimated at $600,000. Dur-

ing the fire's progress, women and children were seen excitedly running in all directions. Many were almost nude, having lost all their possessions in the flames.[87] The *White Pine News* published an extra edition:

> The only redeeming circumstances which will give a crumb of comfort amid the surrounding desolation is the spirit displayed by the people of all degrees, expressed in hopeful and assuring words of comfort to each other. There were so many acts of individual sacrifice of their own property, to assist others more in need, that our estimate of human nature has been considerably raised since half past five this morning. On every hand offers of assistance, from one to another, could be heard, and all seemed to vie in offering every comfort to those in need.[88]

The fire had been set by Alexander Cohn in the back of his tobacco shop on Main Street, near the center of the town, just before daybreak. His business was not thriving, and he hoped to collect the insurance; to assure the success of his plan he had turned off the stopcock in the main water pipe leading into town, thus shutting off the entire supply. Before the cause of the difficulty was discovered, the town was razed. The shopowner was arraigned, convicted of arson, and sentenced to seven years in the state prison. Hamilton was soon rebuilt, but when in June, 1874, the main business portion of Treasure City was burned, little of it was replaced.[89]

THE SEARCH FOR CAPITAL

In the midst of the 1869 summer mining season, the White Pine County tax assessor's office released a report that made somber reading. According to his figures, only 4,174 tons of ore had been milled during April, May, and June with an aggregate bullion yield of $412,814, or an average of only $98.82 a ton. These returns had been compiled from the records of thirty-four mines.[90] During August the weekly shipments of bullion by express did not rise above the $60,000 to $65,000 averaged during June and the month's total did not exceed $300,000.[91] The midsummer interest of San Francisco therefore collapsed. During the height of the summer's boom at Treasure Hill in anticipation of increased silver ore production the erection of stamp mills had been overdone. The number of stamps was far out of proportion to the quantity of ore known to exist, and many lay idle in the fall.

To counterbalance the pessimistic reports on silver production by the tax collector and the express companies, the *White Pine News* pointed out that under the law $18 per ton of gross proceeds were exempt from taxation and were not included in the assessor's calculations. The editor gathered information from all the mills in eastern Nevada handling White Pine ore in an attempt to prepare a more optimistic statement on the district's production. Total shipments for the ten previous months from Shermantown, Treasure City, and Newark were only $765,324. Information from Hamilton mills was not immediately available but it was known to be much the largest because most of the mills in the district were located conveniently close to the stage depot there. The editor hazarded a guess of $2 million or slightly more, for the previous twelve months.[92] Hamilton's newspaper, on the other hand, noted that $1,511,891 had been sent from that town between September 15, 1868 and September 15, 1869, $526,891 from Treasure City, and $352,829 from Shermantown making a total of $2,390,918. White Pine ore handled by outlying mills scattered throughout eastern Nevada might well bring the total to $3 million.[93]

In spite of the evidence collected and the speculation by the journalists, the production during the last three months of 1869 must have fallen far short of that for the same period in 1868. When the date for reporting shifted from September, 1869 to January, 1870, Wells, Fargo reduced the value of bullion shipped from the three communities for the twelve months immediately preceding from $2,390,918 to $1,938,828. Of this amount, the bulk was milled in Hamilton — $1,176,897. A check of destination points revealed that over half of the silver shipped from Hamilton was headed toward the East — $604,282. However, the mills in Shermantown and those shipping from Treasure City were owned by San Francisco capital so their total shipment headed West, making the preponderant trend in that direction.[94]

Although the express company and forwarding agents claimed to have handled approximately $2 million during 1869, the tax assessor of the county collected on bullion valued at only $1,822,867 in that year.[95] The express company figures were doubtless more reliable than the tax rolls, but both probably underestimated the total wealth because many small quantities of bullion continued to be obtained by parties who escaped taxation and who did not ship their silver by express. Mill production figures, a more accurate indication of production, totaled well over $3 million for the year. The Oasis mill alone had processed bullion worth $1,500,000 in coin by September, 1869. When the estimated total production of $1,500,000 silver in the district for 1868 is added to that of 1869, the grand total up to January, 1870 must have been somewhere near $4,500,000.

One statistician was interested in annual production of bullion per man in the district during 1869. One thousand men employed in running the 179 stamps in the district had produced around $2 million according to his estimates, or about $2,000 a man. At the prevailing wage rate of $4 a day, this did not leave much margin for profits. However, the *White Pine News* challenged his figures pointing out that only twenty-eight stamps had been running in the district at the beginning of 1869 with less than ten workers. Revised estimates and calculations produced a more optimistic figure of $3,832 as the annual product of the labor of each mill worker. After deducting $1,460 for his wages, a paper profit of $2,372 from his labor appeared to have been made.[96]

Gross production for 1870 exceeded a million and a half dollars, bringing the total for three seasons to a minimum of $6 million, but the decline from the peak-production year of 1869 had already begun. This record was in no way comparable to that of the first three boom years at Washoe, in Idaho, or Montana, but it compared favorably with the previous Nevada rushes at Esmeralda and Humboldt, and was somewhat greater than the wealth obtained from any of the Arizona rushes. Calculations of the amount of ore produced in any mining district are always inconclusive, and White Pine was no exception. All figures are estimates, but the general trends of production can be easily ascertained.

Although San Francisco capital had supported the development at the Washoe mines and was consistently, if cautiously, moving into the Austin and Eureka districts, Treasure Hill failed to get adequate aid from the Californians. White Pine residents began to complain loudly that "outside influences" were undermining the district and to look elsewhere for financial support for drilling equipment, and milling and smelting machinery. The Central Pacific Railroad was indicted for its exhorbitant railroad tariffs that necessitated excessive charges for equipment and merchandise. Goods were brought over the Sierra Nevada by rail to Reno at three cents a pound, while the same freight delivered to Hamilton cost twenty to twenty-five cents a pound. The railroad charged $46.50 from Sacramento to Elko but only $50 to Promontory Point, more than two hundred miles beyond. San Francisco merchants had a habit of "marking up" goods headed for White Pine and demanding "C.O.D." One journalist, awaiting completion of the transcontinental railroad, remarked, "We want to cut loose from San Francisco and the Central Pacific. We look to the East."[97] By August, managers of some of the leading mines that regularly sent bullion to their company offices at the Bay complained that they were not provided with sufficient coin to pay their men. "It is gratifying, however," noted the *News,* "to observe that we are beginning to feel the effects of Eastern capital seeking investment here." A group of Chicago investors had purchased several claims that needed development; New Yorkers had instructed one of their Nevada agents to purchase a mill site. Once the railroad was completed, several parties of midwestern capitalists came to the district to tour the diggings. The newspaper prophesied:

> When a large share of our mines are owned by Eastern capitalists, as in time they will be, such financial distress as now prevails in San Francisco will cease to affect us seriously. Our Pacific Metropolis has never afforded the mining regions the amount of capital needed for their development, nor proved a sure reliance for carrying through the mining enterprises initiated by its aid. The railroad having connected us with the greater financial centers of the older States, we may in the future depend upon a more ample, stable and reliable money market.[98]

The Chicago *Tribune* announced in October that the National Bank of Commerce had received from White Pine a consignment of silver bullion worth $10,000, forwarded by the Treasure Hill Silver Mining and Milling Company. This was the second consignment received by this bank and others hoped to share in the weekly bullion shipments.

> This, it is confidently predicted, is the commencement, in Chicago, of an immensely important branch of commercial operation, the complete opening up of which will result in untold advantage to this city. Heretofore the silver bullion from these districts has been consigned to California, and there exchanged for money and goods; where also has been the market for mining stocks and the headquarters for mining interests generally. This great interest and, and will, it is believed, if the proper steps are taken, avert to Chicago, which would then become the "silver centre" of America.[99]

One Chicagoan had just departed for Hamilton with a ten-stamp mill, and others, who had bought properties in eastern Nevada, contemplated shipment of larger mills westward. A local journalist commented on this report by suggesting that "the men who invest in and control the mines of White Pine will also control the trade" whether they be from San Francisco, Chicago, St. Louis, or Cincinnati.[100] A hundred or more Chicago men visited White Pine during the fall of 1869. The first group were bona fide capitalists who were "tendered the hospitality" of the mining community. A second crowd was composed of "adventurers and bummers who enjoyed our hospitality without the tender ever having been made." Most visitors left Hamilton impressed by the potentiality of the mines The Chicago *Times* now joined the *Tribune* in urging Chicagoans to establish trade connections with eastern Nevada. The editor of *The Inland Empire* reiterated his position.

> Only a little of our silver stream, as yet, flows that way. It will depend upon Chicago herself whether the main volume tends

thitherward hereafter or not. One Chicago company are here building a mill. They will add to the stream. Whether any more such companies will be formed or not remains to be seen. We have only this to add, in conclusion: There will for many years be thousands of people here to be fed. . . . Some place will get rich in supplying us. Whether that place is to be Chicago or not we don't know, nor do we care; but that it is worth the attention of Chicago, or any other place, we are very sure.[101]

An observant White Pine resident noted that merchants from Chicago were turning their attention to the silver district and were entering into competition with San Francisco capitalists. In his opinion an extensive and profitable business was likely to develop with the Midwest because the money market was so stringent both in California and western Nevada. As an eternal optimist who had spent the last twenty years in California mining camps, he thought White Pine was a country where well-directed industry and judiciously invested capital would harvest great dividends. In his own mind he was sure that the silver mines, or at least the best of them, would be as permanent as those in any other part of the world.[102]

Early in 1870, a San Francisco writer suggested that the Bay City had received more injury than benefit from the discovery of silver at White Pine. The mines, he claimed, had drawn away a great deal more capital than had found its way back. Processing the ore had not called for the purchase of a large amount of machinery. Most important of all, the first excitement resulting from the brilliancy of the prospects had distracted the attention of mining men from more modest California enterprises that in the long run would have proved more remunerative. The editor of the *News* took vigorous exception to this viewpoint, stating that San Franciscans had developed the theme of money being drained "to the interior" ever since the railroad had been inaugurated. San Francisco had received $1,225,000 in bullion from White Pine. In addition she had made a profit of $250,000 in trade. This income had been offset, to be sure, by $1,500,000 invested in mining and milling properties, but many of these were worth more than had been paid for them. San Francisco property in White Pine was appraised at close to three million dollars, undoubtedly less than had been lost in speculations with mining shares.[103]

Some of the first discoveries began to change hands early in 1870. The Hidden Treasure was sold in January for $200,000 to G. E. Roberts and Company. By the end of the year, most of the mines producing valuable ore were no longer controlled by individuals or by partnerships. Syndicates with investment capital now dominated the district. The Bank of California had hastily purchased many promising claims hoping to direct the management of the most productive mines as it had done on the Comstock lode, but these bankers never gained control at White Pine because of vigorous competition of other capitalists.[104] One report claimed there were one hundred and ninety-five incorporated companies in the White Pine District in 1870; another reported one hundred and ninety-seven, with a nominal capital of $277,564,000.[105] The interest of investors and speculators seemed unquenchable. With the opening of the new working season in April, 1870, all types of White Pine stocks began to appreciate on the San Francisco market. Among the operators "on the street" there was considerable buying and selling, and "the prevailing impression appears to be that White Pine is about entering upon a season of unparalleled prosperity."[106]

THE TURNING POINT

The routine life of the community had been suddenly interrupted at mid-afternoon on October 7, 1869, by the news that the Hamilton branch of the First National Bank of Nevada had closed its doors. The San Francisco office had wired its affiliate in White Pine early in the day for $5,000 coin. Wells, Fargo & Co. officials in Hamilton, at first hesitant, finally agreed to authorize the transfer by telegraph. An answer came back from Wells, Fargo in San Francisco, "Paid — but cash no drafts for them — suspended here." The news spread rapidly, panic ensued, and a run was started on both the Hamilton and Treasure City branch banks. A few prompt people in Treasure City had the coolness and the nerve to force their way through the back door of the bank and withdraw their funds before the sheriff arrived.

At midnight, groups of people were still gathered in the streets and public houses, engaged in earnest conversation. Among countless rumors was the report that citizens had seized the Austin branch of the bank and were forcefully taking their deposits. A buggy loaded with coin was said to have left the Hamilton agency early in the morning and parties had departed on horseback to recover the treasure. In time the failure of the First National was attributed to the gold speculations of a San Francisco official in the New York stock market. The traveling agency manager arrived in Hamilton the day following the crisis and, after a hasty examination of the books, reported that only $200,000 had been loaned in White Pine County although the bank had assets valued at more than $300,000. Within thirty-six hours the credulous had filed twenty-one attachment suits against the bank for a total of $60,000. Bets were being taken in the streets of Treasure City that the bank would open in ten days. On the other hand, street brokers were offering twenty-five cents on the dollar for demands against the bank, and in a few instances were finding takers.[107]

In an attempt to ease the community's tension, the Hamilton newspaper reported the following incident:

Yesterday we strolled into a place of public resort on Main Street. Standing by the counter was a lady dressed in deep mourning, with that pensive and melancholy look which in a beautiful woman is touching and eloquent. . . . We were admiring this picture slyly . . . when suddenly the lady roused herself from a fit of abstraction and turning around with a sleep-walkers' look in her eyes exclaimed, "They are d — d swindling thieves, every son of a b — h of them." We left suddenly, sadder and wiser. We understand the lady (you may know she was a lady by the remark she made) had deposits in the National Bank, Hamilton Agency.[108]

Reports circulated that the Bank of California was also shaky, but the officers, anticipating a run, had wired San Francisco for coin, and when depositors lined up demanding payment, all were promptly satisfied. There was no sign of the resumption of business by the First National Bank, however, and White Pine businessmen quickly concluded that the concern was a hopeless wreck.[109] Toward the end of the month a new rumor gained credence that one of the San Francisco directors of the Bank of California had declared bankruptcy with liabilities of $3 million and that the bank would be forced to suspend operations. From mid-morning to mid-afternoon, White Piners once again stormed the doors of the Hamilton branch and more than $100,000 in cash was paid out by tellers before the throng subsided. On Treasure Hill it was payday and miners raced one another from the paymasters' offices at the mines to the bank, checks in hand, hoping to arrive before the doors closed for the last time.[110] Although the Bank of California proved solvent, White Pine residents lost several hundred thousands in the National Bank; its failure marked a turning point in the economic development of the mining district.

By November, complaints were being registered about the stringency of the money market and the depression in business. Mining continued profitable, however, and none of the leading mining companies had announced plans to shut down for the winter.[111] At the end of the month, cold winds forced a suspension of all outdoor work on the mines, but the snow was not yet deep enough to retard the shipment of ore. Business became very dull by mid-December when the company payrolls put between $75,000 and $100,000 in circulation to provide momentary relief.[112] Immediately after paying its employees on December 15, the Chloride Flat Mining Company shut down all work on its mines. Other companies announced similar plans and the ranks of the unemployed were rapidly increased. As usual, mine owners were accused of deliberately attempting to depreciate the value of White Pine stocks.[113]

By the time the new year opened, the leading mines and mills had all announced plans to close for a month or two. The sanguine insisted that the depression would not last longer than six weeks, that all business would receive new life in the springtime with the renewal of immigration from California and the East.[114] The Hamilton editor chided the community for its impatience at the failure of capital to flow into the district in accordance with its capacity to produce. Self-interest dictated that the owners of claims should continue to work them even though the mills had used the inclement weather as an excuse to shut down.

> To the men of White Pine, who give way to despondency, we would say, "You have a long and rigid Winter to worry through. Whatever you can do in the way of prospecting or developing, in the meantime, will not be lost. If you can get your diurnal ration of pork and beans, thank God, and think that it is the precursor of boned turkey and cranberry sauce. Your mines are the best property you could have. Labor spent upon them will be the most remunerative labor you ever performed; for capital is opening its eyes to our magnificent possessions, and will before another season has passed, beg humbly for a piece of our great silver pie."[115]

Few took the journalist's advice, for in February, 1870, he reported that crowds of men who were owners of good claims were walking the streets in idleness and complaining about hard times. Such men were reminded that it required about as much bread, meat, potatoes, and onions to support an individual whether he worked or not. "Industry often meets with a reward, while idleness is generally crowned with poverty," he prophesied.[116]

Real estate values fell to the lowest point since the discovery of silver. Local residents protested that while the price of real estate had gone down with the receding tide of population, the bullion product of the mines had increased just as steadily.[117] Businessmen were urged to demonstrate enough public spirit to secure the opening of all roads leading into Hamilton without toll. Mill owners in the outlying mining districts of eastern Nevada had to come to the foundry in Hamilton for their castings, and with adequate, non-toll roads they might elect to purchase supplies, thereby making the county seat the business center of the region.[118]

———•—■—•———

The serious side of the newspaper war was recognized at the close of the mining season of 1869 when the *White Pine News* announced it would suspend daily publication beginning in September and would become a tri-weekly, appearing on Tuesday, Thursday, and Saturday mornings. Daily publication had become unprofitable and until there was a general improvement in business, anticipated for the following spring, it would not be resumed.[119] This pioneer newspaper had been in difficulty for some time. In May, 1869, W. H. Pitchford sold out and turned the journal over to W. J. Forbes and Company. Early in August the enterprise was incorporated as the Daily News Company, capitalized at $50,000 divided into 5,000 shares of ten dollars each. Forbes, John I. Ginn, a staff writer, and W. H. Pitchford, Jr. were joined by two other White Piners, John Church and R. L. Tilden, as trustees. By the month's close the White Pine News Printing Company was listed as publisher, and prices had been reduced from sixteen to twelve dollars a year.[120] In these same weeks the proprietors of *The Inland Empire* informed patrons that their present office was too limited for business requirements and that the publication would be removed to the second story of a stone building at the corner of Main and Silver streets in Hamilton.[121]

When winter came the *News* reverted to a weekly, and the editor candidly explained:

> All other business is drawing to a shortened expense for Winter and we must conform to the custom. While mining is improved, and additional milling facilities are swelling the bullion product, trade in general seems paralyzed. Our business men can not afford to advertise; and so, we conclude that a weekly issue will meet the newspaper requirements of the community until times limber up.[122]

The Inland Empire soon announced that it also was "taking in sail" to wait the approaching storms of winter. Although continuing as a daily, the pages would be reduced to original size. The reason was hard times. At last the publishers of the two newspapers were in agreement: "At present the men who are making money fast in White Pine are the mill owners and the proprietors of mines. Both of these classes have no interest in a newspaper further than to glean its news at an expense of seventy-five cents a week. They rarely have any use for the advertising columns of a journal." The editor hoped his paper could "spread full canvas to the breeze again" when the next season opened.[123]

The owners of the *News* had by no means given up the struggle. The *Omaha Herald* reported Ginn's purchase of a new press and an entire printing outfit to be transported to the White Pine district "where he will set the Republican ball in motion in the great silver region as soon as he can get the material freighted through." *The Inland Empire* laughingly reported the account. As the sniping continued during the winter, the Hamilton paper remarked in bitter sarcasm, "We sincerely hope the *News* editor's affairs may speedily improve — for some men are so constituted that when times are hard they can neither be good natured nor candid in their statements."[124]

In December rumors were afloat that the *White Pine News* was planning a move from Treasure City to Hamilton when its Omaha press arrived. *The Inland Empire,* contrary to its announcement of the previous month, revealed that pressure on its advertising columns compelled it to go back to the larger edition. The *News* was ominously silent, but a week later rumors were confirmed when it announced a "change of base." The removal to Hamilton was imperative because that town had become the center of trade and government. The publishers frankly admitted, "To be on an even footing for a division of legal and other advertising we are constrained to place our facilities in that respect within easier reach of patrons." The primary reason was "vigilance in our own behalf as business men."[125] The struggle for survival now began in earnest.

In mid-winter, *The Inland Empire* revealed that G. A. Brier, a journalist from Virginia City, had bought out the senior partner of the paper, J. J. Ayers, and planned to renew publication. The *News* immediately, on January 29, 1870, announced resumption of a daily issue. Both newspapers competed for subscribers from their Hamilton offices, and within two months the *News* appeared to be getting the upper hand on its competitor, in spite of the fact that *The Inland Empire* had published longer in Hamilton. Even in the midst of dull times, when everyone was economizing, the list of subscribers to the *News* increased and the press ran overtime.[126] Early in April, the *News* announced that it was to receive national and foreign events by telegraph since its increased circulation justified the expenditure of an additional $300 a month for this service. The material appearing daily in *The Inland Empire* under the head of "Telegraphic" was now copied from the *News* of the previous day. Two weeks later the *Empire* ceased publication.[127]

Shermantown had been without a newspaper since the *Evening Telegraph* had folded. With the aid of A. Skillman, G. A. Brier, who had edited *The Inland Empire* in the months prior to its demise, made plans to revive the defunct *Evening Reporter.* The first issue, under Brier's editorship, appeared on May 9, 1870, and the editor of the *News* congratulated his former rival for "much original and very little *cut* matter in the news columns."[128] Ten days later Brier dropped dead in the office of Wells, Fargo in Shermantown. Shortly before his death, he had purchased quinine in a local drugstore. Upon swallowing the potion, he showed symptoms of spasms and cried out, "I am poisoned." A bitter controversy ensued as to whether or not a pharmacist had mistakenly given him a deadly dose. Brier, a native of Indiana, died a bachelor of forty-one, leaving a brother in Kansas and many friends throughout California, in Virginia City, Gold Hill, and Meadow Lake, Nevada, where he had worked as a journalist before coming to White Pine. The *News* admitted that although "he had his social eccentricities," he "possessed many noble traits of character."[129] Soon thereafter the Hamilton paper noted:

> For the past two days we have missed our little evening visitor from Shermantown. We are now assured that The Reporter is no more. It collapsed of inanition in the business department, and has gone the way of all poorly-sustained newspapers. The late numbers were written up in a lively and spirited vein, and the mechanical department was creditable. . . . The rumor that Chicago capitalists are about to resurrect it cannot be traced beyond the proprietors themselves.[130]

Shortly the press was moved by Skillman & Co. to Eureka and there used to start the *Sentinel*.[131]

The *White Pine News,* first newspaper in White Pine, had now survived all competition, but the victory was an empty one. The editor soon was admitting, "We feel like apologizing to our readers for the scarcity of local matter in this issue. To be candid, but little transpired; besides, we are mentally and physically sick — generally demoralized."[132]

———— •—• ————

Many poorer but wiser claimants had left the district during the summer of 1869 when it became apparent that purchasers for their claims were non-existent and capitalists had no intention of making large investments. In the fall when improved methods of smelting base metal were introduced, some of their claims became valuable. By the spring of 1870, claim jumping had become a systematized business. Many jumpers went to extracting ore within eight to ten months after the owners left White Pine, without waiting for the year's time-limit to expire. One smelting firm employed agents to search the records in the recorder's office for claims that had elapsed and proceeded to gobble them up. Violations of the time-limit were certain to lead to lively litigation, and many of the lawyers who had left the district during the winter of 1869–70 were back on the scene by mid-March, and the *News* commented:

> Although we are now blessed with a bar which is not excelled in ability or surpassed in numbers by any district of the same size on the Pacific Coast, we look for a largely increased display of "shingles" on our new vacant houses in the course of a few months. The old prospectors are beginning to come back to find their

claims in the possession of other parties, and are waxing wroth. They swear worse than the army did in Flanders, and pistols or law seem to be uppermost in their brain. There may be a slight resort to the former; but we think the lawyers will in the end carry off most of the spoils.[133]

Soon the press was complaining about the alacrity with which miners appealed to the court. Everyone now recognized that slipshod mining laws of the district imperiled all claims, but only cases of a palpably fraudulent character could be settled by the court. Miners were again advised to avoid cases where the rights of each party were faintly defined, but to settle differences by consolidation or arbitration. Investors were being deterred by the incessant mining litigation that advertised the difficulty of securing an absolute and undisputed title in White Pine. Unscrupulous jumpers were declared public enemies comparable to those who would steal a horse or rob a stage.[134] Either because the majority of claims were not worth fighting for or because those who engaged in a court battle too often in the past had emerged impoverished, the miners finally heeded the advice of the journalists. When the federal district court opened its July, 1870 session, the paucity of suits brought forth complaints from the legal fraternity. One newspaper noted,

> There is scarcely a case apiece for the entire bar. And that is not the worst of it. We doubt, from an examination of the docket, whether there is a "piece" a case for the lawyers. There must be something done. If people will not go more generally into litigation here, we may lose a large slice of our present extensive forensic talent.[135]

Many White Piners had looked forward to the construction of a railroad to replace stage and freighting lines thereby giving greater permanency to the community. As early as the spring of 1869 there was talk about a Central Pacific Railroad plan to build a branch southward from the main line at Elko. As soon as the transcontinental was completed, it was predicted the company would grade the right-of-way in two months' time with the track and iron horse following a month later. More optimistic residents insisted that a railroad was not only a pressing necessity but also a profitable investment for those undertaking it. Two groups of promoters apparently shared their enthusiasm for connecting the Central Pacific Railroad with the Colorado River and the proposed southern transcontinental. An Austin company was organized to build from Elko to a point opposite Mojave City, on the Colorado; a more powerful San Francisco group, led by Frank Drake and J. W. Crawford, had a similar scheme. Both companies, not anticipating that the line could pay its way, sought land grants and subsidies from the federal government.[136] Meanwhile, the state legislature of Nevada approved an election in White Pine, Elko, Nye, and Lincoln counties to authorize county bonds for the support of a railroad running from Elko to the Colorado River. Each of the counties had from thirty to fifty miles of line within their boundaries and were authorized to subsidize construction accordingly from $150,000 to $300,000. To get authorization for this bond election the Colorado and Nevada Railroad and Navigation Company, that planned construction without a land grant, agreed to pay the election expenses in any county if the majority of the residents disapproved of the issuance of the bonds.[137]

The railroad dream of 1869 did not materialize, but in March, 1870 Nevada's senators, Stewart and Nye, introduced a bill in Congress granting public lands for a branch line from the Central Pacific. The *White Pine News* immediately objected by saying, "We want a railroad from the Central to the Colorado; and we will have it, in time; but we do not want it so badly as to ask Congress to allot to the construction company the bulk of the available agricultural lands in this State, to induce the building." The editor thought it an outrage that the senators proposed a subsidy to construct a short branch line that could find inducement enough in the growing business if the country opened up. The newspaper called for the retirement of both senators if they pressed for this bill.[138] Just before its demise, *The Inland Empire* joined the fight insisting that the government aid was indirectly destined to go to the Central Pacific corporation.

> We trust that no further Government subsidy whatever shall ever find its way into the coffers of this closed corporation, which is managed by six or eight individuals who have given all their contracts to their own "ring," so that they could fare sumptuously every day; build palatial residences for themselves; water their stock; bid defiance to the men of small means who had invested their money in it when it could not have been constructed for the first thirty miles without their aid; charge a cent or two a pound more from San Francisco to Elko than is charged for two hundred miles further transportation, and defy the power of States and of the Nation to interfere with its passenger or freight tariffs.[139]

Two months later the *White Pine News* resumed its campaign to interest capital in railroad construction as a business venture. To move the estimated summer freight of one hundred tons of ore and bullion daily, one hundred sixteen-mule teams would have to be continuously on the road, and since only a few would have returning freight, transportation costs would be fifty dollars a ton. The *News* insisted, "The question is, shall we depend upon the slow, uncertain, and expensive method of having 100 sixteen-mule teams and lose eight days every trip, or shall we have a railroad which will take it in as many hours, and for one-tenth the money?" The valley between Elko and Hamilton would not require expensive grading for a line, and any company of enterprising citizens undertaking the task could possibly prevail upon the Central Pacific to provide the ties and rails as a means of encouraging a feeder line.[140] The newly-organized political clubs, both Democratic and Republican, were urged to sponsor a town meeting to discuss the railroad situation, so vital to the economy of the district. "We must have a railroad," insisted the journalist, "we can not do without one; and the sooner a start is made the better it will be for all concerned."[141] A mass meeting of citizens was assembled to promote a preliminary survey of the route and to obtain estimates of construction cost.[142]

Meanwhile, the House of Representatives was considering legislation to aid the building of an eastern Nevada railroad.[143] In addition to providing government subsidies, this legislation permitted a construction corporation a year in which to commence construction of the Elko-Hamilton line, two years to complete the first ten miles of the road, and one year for each succeeding ten miles. At this rate, the company would be given twelve years to finish the line. The terms were designed to encourage Nevada capitalists, political friends of Congressman Fitch, to undertake the project. The *News* denounced the scheme as treacherous and proclaimed that "the railroad project is betrayed in the house of its friends."[144]

Rather than see Fitch's friends so favored while the laying of the rails was delayed, the newspaper now preferred the legislation proposed by the Nevada senators to encourage immediate construction by the Central Pacific organization. Without blushing, the editor changed his attitude toward the transcontinental combine and announced he was "surprised to learn of one or two men in the community, ordinarily sound thinking men" who were so ardent in their partisan politics as "to enveigh rantingly against the railroad movement, if possibly it might have a connection with the heavy railroad men of California." Such "timorous political panderers" were enjoined to keep quiet. The only hope left was with the builders of the Central Pacific.[145]

> The Central Pacific Company has the material lying idle which would serve to construct and equip the entire road, and probably the best chance for success, in aiming at the speediest completion and opening of the proposed road, lies in the possibility of some arrangement with that Company. A year ago, one of the prominent stockholders in that Company was conferred with as to why they did not put out a branch to concentrate the White Pine trade. He replied that some people designate them a grasping monopoly; that they preferred to make no effort for branch roads unless in

case the people most immediately concerned made a movement
and asked their co-operation; that they thought well of White Pine
and its surroundings, and would at all times be ready to hearken
to propositions from the people.[146]

When the summer working season passed without commencement of the
railroad, even the most hopeful White Piners were bitterly disappointed.

Railroad promoters once again turned to the Nevada state legislature
for support. In the 1871 session, an act was passed to encourage construction of the Eastern Nevada Railroad from Elko to Hamilton by
permitting the counties of Elko and White Pine to issue construction
bonds.[147] The legislation was further liberalized in the 1873 session
authorizing the counties to levy specific taxes not only to pay interest
on the bonds but after five years also to retire the principal by this
method, provided not more than one fifteenth was paid in a single year.
The railroad had to complete the first fifty miles of the proposed thirty-
six-inch-gauge line by July 1, 1874.[148] Even this exceptional generosity
did not induce businessmen to undertake construction, and the railroad
project folded. Eureka appeared to be a more permanent mining center
than Hamilton and a branch rail line was constructed from Palisade
southward to that camp and on to Ruby Hill during 1874–75. Battle
Mountain, on the Central Pacific Railroad, was also connected with
Austin during 1879–80 by the Nevada Central Railroad.[149]

CHAPTER FOUR

RALLY AND COLLAPSE

Nevada mining promoters continuously sought investment capital not only in San Francisco and in the eastern United States but also in Europe. "Colonel" William Sydney O'Connor arrived in London during 1864, before the Civil War in the United States was over, to promote the sale of claims in the Virginia City District. The prospectus of his Washoe United Consolidated Gold and Silver Mining Company, Limited, stated that the wealth of the Washoe mines was "inexhaustible" and that failure was "impossible" if capital was judiciously employed in development. Because of the known wealth of the Comstock, this company was successfully floated.[1] The directors spent British capital lavishly but paid no dividend, and within three years the company was approaching bankruptcy.[2] Before this unfortunate turn of events, however, the "success" of O'Connor's enterprise, and of the Comstock in general, had been used to launch three Anglo-Nevada mining companies in the Reese River area during 1865. Neither promoters nor investors seemed concerned over their ignorance or willful disregard of geography.[3]

In 1867, another mining "Colonel," David Buel, was sent to the French Exposition in Paris as a representative of the Reese River District. The ore samples that he carried with him were certified by authorities of the School of Mines in Paris for an Exposition Medal. From Paris the colonel went to London "with a large collection of rich silver ores from Eastern Nevada, and an almost equal bulk of certificates and recommendations." Noting that Buel was making his presence felt in British financial circles, the *American Journal of Mining* expressed concern over the amount of space devoted to his activities by the London *Mining Journal*. According to reports, the colonel claimed the best Nevada mines produced ore worth $2,000 a ton. In the interest of all

Western mining communities seeking investment of foreign capital, the editor of the New York magazine protested against this flagrant exaggeration. The custom of English financiers who sent a mining engineer to examine mineral deposits prior to purchase was strongly commended. The New York editor knew that Nevada had some good silver mines and "British capitalists cannot do better than to invest money in American mines, nor worse than to make the investment without caution and full knowledge."[4]

Although British investment in the Washoe and Reese River mines was small, when news of the White Pine rush reached London, investors lent a more willing ear. Early in the summer of 1869, a Treasure City resident by the name of Wearne began collecting an assortment of fine ore from the mines of the White Pine District to show capitalists and "scientific men" in Europe, and to exhibit in the leading museums of the United States and abroad. His primary aim was to obtain financial support and technical skill for the construction of an elaborate smelting works in the town of Swansea.[5] In the same month Captain William Boyle of Treasure City sold the Pennsylvania Mine, the first location south of the famous Hidden Treasure, to Colonel O'Connor, now described as one of London's wealthiest men who had "great experience and energy" in mining affairs. Both men left London after the sale, headed for the White Pine district. O'Connor planned to build a twenty-stamp mill on his property.[6] By the winter of 1870, European capitalists were displaying far more enthusiasm for Nevada mines than Americans. Two gentlemen, one English and one German, arrived at Treasure Hill in January with a large amount of capital that they proposed to invest in the smelting business. In the same month an agent of Colonel Buel sold a Eureka mining property to an English company for $300,000.[7]

The investment of British capital in the mining districts of Nevada at the close of the Civil War in the United States was only the initial flow of a major stream of money that was to find its way into the mining kingdom of the trans-Mississippi West. With the possible exception of California, Nevada mines were more popular with the British investing public than those elsewhere in the American West between 1864–73.[8] Within Nevada, the bulk of the capital was concentrated in the central and eastern part of the state, particularly in the Eureka and White Pine districts. Throughout the decade of the 1870's, British mining investments were scattered all over California, Nevada, Colorado, and Utah. Nevada properties continued to be popular. Twice as much money was put into mines in this state as in Colorado.

In February, 1870, a group of London stockbrokers, merchants, and accountants led by E. W. J. Ridsdale, an assayer of the Royal Mint, formed an association known as the Eberhardt and Aurora Mining Company, Limited, with legal authority to purchase lands, mines, and minerals in the state of Nevada and also to buy shares in any other English or American companies owning mines in that state.[9] The next month this company was officially registered with an authorized capital of £500,000 in £10 shares. An option was acquired on the properties and claims of two San Francisco companies, the Eberhardt Mining Company and the Aurora Consolidated Mines. The vendors were Edward Applegarth, Frank Drake, J. W. Crawford, and E. R. Sproul who had obtained and held control of the Eberhardt and North Aurora claims on Treasure Hill. They were to receive £300,000 for their properties, one-half in cash and the remainder represented in the capital structure of the company by paid-up shares. These securities were to be held by the British purchasers until £125,000 in profits had been divided among the shareholders.[10]

Prospective investors were told that the Eberhardt Mine had produced $1,500,000 in coin between June, 1868 and January, 1870, and that the ore yielded from $75 to $600 a ton. All shares placed upon the market were subscribed for in less than two days by approximately three hundred people. Over forty per cent of the stockholders listed their occupation as "gentleman," many of whom were known to be professional investors. The next largest occupation represented on the shareholders' lists was that of stockbroker. The remaining shares were widely scattered among professional people — doctors, ministers, Army officers, solicitors, and accountants. The bulk of investors bought only five to ten shares in the enterprise; less than ten per cent purchased more than a hundred shares. The largest shareholders were stockbrokers, five of whom held over three hundred shares. Included in this select group were the partners in the firm of Haggard, Hale, and Pixley that had brought the American vendors and British investors together and promoted the company. These shares represented, for the most part, their compensation for services rendered.[11]

Sale of the Eberhardt and other claims had been made contingent upon a favorable report by an inspector hired by the English company. Melville Attwood, a mining engineer, arrived in Hamilton by stage on April 4; a week later he was followed by Ed Applegarth, a vendor, and Thomas Phillpotts, a prospective manager, both coming directly from London. These three men, joined by Frank Drake of San Francisco spent a week examining the claims on Treasure Hill, then departed for the Bay city where reports were prepared and forwarded to London in care of Phillpotts. The British representatives estimated that the ore already uncovered at the Eberhardt Mine would produce £500,000.

Upon the receipt of this news in England, capital was called up and the sale consummated.[12] Finally in June sale of the Eberhardt to the British was legally proclaimed, and the Hamilton newspaper reported, "The English Company who have bought the properties will doubtless take possession in a few days, when we may look for an active and extensive working of these first class mines."[13]

When news of the proposed company had first reached White Pine, the press was jubilant. London had led the way. Now that the enterprise was a reality there were reports that the Pogonip, Othello, Tom Paine, and Pocotillo mines were sold in Chicago. San Franciscans were investing heavily in the base metal belt.[14] In truth, the English decision had saved the mining community from economic collapse. Throughout the winter creditors had been lenient, but by the spring of 1870, business had not revived and the hustle and bustle of previous seasons no longer existed.[15] All White Piners now expected the flow of investment capital to bring about a return of "boom times" as a reward for patience and forbearance.

At the first general meeting of the company in London the shareholders' optimism appeared unbounded. Phillpotts, who had now been named manager of the American properties, reported that the two recently-purchased San Francisco companies had earned profits of £48,000 in 1869. The ore had been processed at the Oasis Mill, but with a larger and improved stamp-mill to handle the great volume of available ore, annual profits should reach £96,000. A sum of £50,000 was earmarked for erection of such a mill, raising the initial cash outlay of the British to £200,000. A partner in the firm of Haggard, Hale, and Pixley, commented, "Do not let us build castles in the air, but avail ourselves of Nature's gifts, by applying our engineering application to realize what she has sent us." Larger and more efficient mills were necessary and the sooner they were built the better.[16]

An old resident of Nevada, reading the report of the company meeting in the London *Mining Journal,* suggested that if £96,000 annual profit was to be made from $40 ore, mined and milled at an expense of $20 a ton, it would be necessary to mine 22,000 tons a year from the company's properties. The whole White Pine District had not produced more than 38,000 tons the previous year. The company challenged his statement and reminded shareholders and critics alike that an inspection had been made by two able mining engineers. To this the old White Piner retorted:

> The promoters of this company assume to themselves a virtue of "looking before leaping" — in other words, having the mines inspected prior to purchase. But where, I should like to ask, is the

mining company that has not done the same? Reports by mining engineers, many of them by gentlemen of high standing and unimpeachable integrity, are appended by the dozen to every prospectus. And I appeal to any impartial observer to say whether the promised golden harvest is realized one time in twenty. Somehow or another they do, time and time again, get the wool pulled over their eyes.[17]

The White Pine mining community was disturbed by reports that the Eberhardt and Aurora Company, Limited, intended to import labor from England under contract wages lower than prevailing rates. The local press reminded the company that the scheme had been tried before in California and Nevada and failed to accomplish a reduction in operating costs. Cornish miners, on arrival at the Pacific slope, refused to work for less than the United States citizens on neighboring claims. Moreover, a lowering of wages always was an indication of a want of confidence in the future prosperity of the district. "We have no objection to the English Company making the experiment," observed the *News.* "That is their business. But we feel they will regret it, not only as a failure; but also as a stroke of bad policy."[18] Under these circumstances, an acute restlessness developed among White Pine miners and many decided to try their luck in "far-off fields." The newspaper thought these men had been too easily discouraged and cried out, "Don't go." The editor warned,

> It is just getting out of the fry-pan into the fire. Of what avail is it to the miner to go from this District to another? . . . No matter where we may turn, we will hear the same distressing cry. What, then, is to be gained by the change?[19]

When Beachey, Wines and Company's stage pulled into Hamilton on July 11 the new Eberhardt crowd, including Frank Drake, was aboard.[20] The English promised employment to three hundred men on the Eberhardt, and Phillpotts had apparently returned with unlimited amounts of cash to buy mining claims and town property in Hamilton, Treasure City, and Shermantown. Everyone in need of cash could sell his claim to the British for $100 to $900, with most being purchased at $350. Deeds were filed for the Red Hill and Red Rover claims, the Protection, the Wild Irishman, the Morning Star, and the Idaho mines, and for a series of ledges known as the Eclipse, the Borealis, the Sproul, the St. Louis, and the New York. For twenty-five-foot lots on Main Street in Hamilton, Phillpotts paid $500 although the owners had often obtained them for $50 or less. The old International Saloon building also came into British hands.[21] Business conditions that had been in the doldrums immediately improved.

The British enterprise in Nevada was directed by a large staff, in addition to the general manager, including: George Attwood, an English engineer, as milling superintendent; James Pott, chief foreman of the mines on Treasure Hill; John McChrystal, foreman at the North Aurora; and John Goodfellow, foreman at the Eberhardt.[22] In London, company directors worked out an elaborate reporting procedure on the operation of the mines to be made by each employee in Nevada. The manager was expected to send weekly reports on the amount of bullion received, the cash disbursed, and a letter summarizing general operations. He was also to certify the monthly reports submitted by each mine and mill superintendent and to make a private, confidential report on the activities in each department. Mine superintendents listed the number of men they had in their employ each week, the wages paid, the output of ore, cost of the output, the number of fathoms sunk, the quantity of ore sorted and sent to the mills, and the assay records. In turn, the mill manager was to estimate the amount of ore on the dump, the quantities received from the mines weekly, the number of hours the mill machinery was run, the amount of bar silver produced, the quantity of mercury used to recover the silver, and a wage sheet showing the number of men and how employed. The secretary-accountant was expected to provide a weekly balance sheet, including a statement of all stores purchased and distributed, salaries and charges paid. Copies of his cash book, journal, and ledger were to be sent to London every two weeks and the monthly bank statement was to be countersigned by both the general manager and an agent of the bank and then forwarded to England. To assist the secretary, the company paid a subordinate officer known as the storekeeper. The directors notified Phillpotts that this regular checking system must be carried out. "The whole system of accounts is thoroughly effectual and to exercise a complete supervision of every receipt and disbursement and any and all other information in detail," suggested the company chairman.[23] If these reporting procedures had been rigidly adhered to by the company's representatives in Nevada, keeping records and accounts would have consumed more time than milling and mining operations. However, the directors were more successful than those of most Anglo-American mining enterprises in the American West in getting regular reports from their managers and agents.

Phillpotts, staggering under the responsibility and labor required by the London office, notified the directors, "With regard to my own remuneration, I will say nothing until my return which will be as soon as I possibly can for altho I never contemplated staying out here so long and having such important work to do, I feel in honor bound to carry out what I have promised." In August, one director reported that the

Board had no doubt that "perfectly satisfactory arrangements" would be made concerning his remuneration and that, in the meantime, his exertions were appreciated. Another wrote, "I do not think your services will go unrequited. We have said little and we have thought all the more." As the summer working season of 1870 came to a close, Phillpotts made arrangements to return to London. The directors authorized his trip but also noted, "We trust that you will not think of leaving for England until all is in order and you feel quite certain that matters can progress in your absence."[24] Phillpotts was the brother of one of the largest stockholders in the company; some of the directors were from the first concerned over nepotism and others had developed reservations about his mining knowledge and objected to his extravagance. From the beginning the Board appeared to be sharply divided in its attitude toward the manager.

In spite of the auspicious beginning of the English enterprise, the White Pine District remained demoralized by the failure to uncover more silver ore. An exodus to Eureka and Pioche was under way by midsummer. Few of the mills remained in operation and no furnaces were running.[25] The bullion produced from January 1 to July, 1870 was valued at only $329,251, with lead and copper accounting for a larger percentage of the profit than silver. Furnace and mill men looked upon inquiries about production as impertinences and a few refused to report, providing thereby the feeble excuse that the poor showing was the result of inadequate statistics.[26]

In the formation of this joint-stock company to invest at Treasure Hill, the British had followed customary procedures in promotion and registration. A temporary association of interested British investors had been legally formed, with wide latitude of activities, to seek a suitable property. The sale of White Pine mining claims in London appears to have been conducted by vendors with a minimum of interference by stockbrokers and go-betweens. As usual, the transfer of the property to the British company was contingent upon the report of mining experts who were to check upon the representations made by the American owners. When these reports proved favorable, they were used for public consumption, being quoted in the prospectus, in handbills, and in newspaper advertisements attempting to sell shares. The occupational and professional pursuits of the investors, as well as the size of their holdings, were typical of the American mining companies being floated in London at this time. Like most such enterprises, this British company in White Pine was overcapitalized, but unlike the majority of foreign mining enterprises in the West, this concern had sufficient funds for extensive development work. British directors decided that a manager sent from

England to Nevada would be more loyal and responsive to their interests than an American superintendent. Although they experienced difficulty in the selection of personnel, the London officials had demonstrated both concern and competence in handling the problems of long-distance management by drawing up an elaborate set of instructions for each of the employees concerning reports to the London headquarters.

At the height of the boom at Treasure Hill, residents had cried out for capital to sustain the expanding economy. San Franciscans had rushed in momentarily, hesitated, and withdrawn. Easterners from Chicago and elsewhere invested moderately. Now from London more capital had arrived than the most sanguine Nevada residents had hoped for. The test could be made. Was investment capital all that the White Pine District needed for phenomenal prosperity?

BRITISH CONTRIBUTIONS

The Eberhardt and Aurora Company continued to pour capital into the White Pine Mining District in the last half of 1870. The directors were convinced that profits from ore reserves could be increased by the construction of a larger and more efficient processing plant. C. W. Lighter, an English engineer, was dispatched to survey the situation. Phillpotts informed him of stage schedules between Elko and Hamilton and added, "I need scarcely to advise you to be cautious and reticent, not knowing anything as to the intentions of the new Company, for as you may well imagine, curiosity is about here."[27] Shortly after Lighter's arrival, Phillpotts went back to England on business and Edward Applegarth, one of the vendors, returned from London to White Pine. There Applegarth learned for the first time that the company planned to build a new mill, and "he was very much excited, and thought he had not been treated with common courtesy" in not being informed in Britain. He threatened to telegraph the London office and Phillpotts as well, but George Attwood advised him to be patient and await the manager's return.[28] Lighter came to the conclusion that the Oasis Mill was faulty. Ore from a pocket of exceedingly rich chloride of silver had at first yielded bullion averaging one hundred dollars a ton, but shortly thereafter production sharply declined. The mine manager insisted that the same rich ore was being shipped to the mill, and that the precious metal was being lost in the processing.[29] Speedy construction of the new mill seemed imperative. Both Frank Drake and Applegarth were disturbed about the criticism of the Oasis machinery and feared that the English company might demand a financial reconsideration of the sale. Attwood wrote Phillpotts, "Drake is I think a little hurt about the mill business, but I talked very strongly to him on the subject and told him how you had done all for his advantage as well as the rest of the company. Also I showed him that you had been the best friend that he ever had, and he says, *its so.*"[30]

In September the company announced that the International Mill, of sixty stamps, was to be constructed and that it was "intended to be the most complete in all its details that has ever been erected, if money and practical knowledge, gained by long experience . . . can make it so." The machinery had been made by Booth and Company of San Francisco, and the California papers reported that "upon trial the engines were found to be not only the most complete but the best working engines ever built on the Pacific Coast." To expedite construction, timber was to be hauled in from Truckee, already fitted and morticed.[31]

The activities of the English company at Eberhardt City, site of the new mill, was making that town "the liveliest part of the district." Early in October the press reported that a large boiler had passed through Hamilton for the new Eberhardt mill. "We understand that there are fourteen large teams en route from Elko, loaded with machinery for the same company," remarked the editor. By October 15, grading for the "mammoth institution" was completed and as fast as the lumber and equipment arrived, the carpenters and machinists were being put to work. The next month twelve wagons, three of which were drawn by nine yoke of oxen, arrived in Eberhardt. Large quantities of timber, crushing machinery, and engines were on the ground, and local citizens observed, "The substantial masonry and thorough construction of the mechanical parts and buildings indicate the confidence of the builders and insure success."[32] George Attwood revealed that "this masterpiece of mechanical skill will be in working order by the commencement of December [1870]."[33] Working forces at both the mine and mill increased daily and the wages paid by the English helped to revive the sagging economy of the community.

When Melville Attwood was in White Pine inspecting the Eberhardt and North Aurora, he also made an extensive examination of the South Aurora Mine, adjoining the Eberhardt, in the interest of another syndicate. His report was not only favorable, it was also enthusiastic.[34] As a result a second British enterprise, known as the South Aurora Silver Mining Company, Limited, was registered in 1870 with an authorized capital of £300,000. This company paid two-thirds of the £300,000 purchase price in cash, rather than the half paid by the earlier London company. Melville Attwood told the promoters that this mine had sold silver worth £73,000 in London since January 1, 1869, and predicted that at the current rate of production, dividends would be twenty-five per cent a year. If a projected expansion of mining and milling operations proved successful, shareholders might reasonably expect anywhere from sixty-five to seventy-five per cent a year, according to his calculations.[35] Among the directors of this company were Leland and Asa P. Stanford of San Francisco, directors of the Central Pacific Railway and owners of the Stanford Mill near Treasure City.[36] The promoters agreed to establish a committee in San Francisco and a Board in London with the understanding that if more than one-half of the company capital was subscribed in Europe the London group would control the management of the company. If United States citizens put up most of the money, the directing Board would operate from San Francisco with a subcommittee in London. Each director or committeeman had to subscribe £500 to the enterprise.[37] Capitalization could be increased to £400,000 to allow the public to invest. As the sale of shares progressed, Melville Attwood reported the discovery of a rich body of ore in the center of the South Aurora chamber that produced from $900 to $1,000 of silver a ton. A single sample ran $1,254. In the first two weeks of September, $33,040 in bullion was shipped to London.[38]

In spite of such favorable reports, the sale of shares moved slowly. Investors, widely spread throughout England, engaged in over forty different professions and businesses. The largest holders were Isaac and Leopold Seligman, British brokers, each of whom had been given approximately ten thousand shares as payment for floating the company. Four merchants held from five hundred to fourteen hundred shares each, twenty-four individuals subscribed for from one hundred to two hundred shares, but the great majority of the remaining three hundred and forty subscribers owned less than twenty shares.[39]

While the British planned heavy investments in White Pine, American-sponsored mills and furnaces in the district were being forced into bankruptcy. In September, 1870, the White Pine Smelting Company in Swansea had failed and settled its obligations for thirty to fifty cents on the dollar. A reorganization was promoted in New York; the furnace was refired for a few weeks but shortly the sheriff placed the works on the auction block. Rothchild's Works of Hamilton, erected by Chicago capitalists at a cost of $60,000, was under attachment for charcoal, labor, and even the butcher's and grocer's bills. Some of the companies, including Rothchild's, had accumulated large stocks of silver-lead bars that had been shipped to San Francisco for further refining, and the smelter owners followed, in turn, to obtain the precious metal, leaving the miners and operatives to "whistle for their pay." Only twenty-five men remained at work on their claims in the Base Metal Range in September; less than that were on the spurs of Treasure Hill. Only one smelting company was in operation and the proprietors insisted that each miner must pay the cost of delivering his ore to the furnace. Managers then assayed samples and placed an evaluation on the various lots brought in. If the miner was not satisfied he had no alternative but to cart away his ore at his own expense. Men bringing in ore for which they expected to receive $20 to $30 a ton were often offered only $8. Prices were so low no man could live by his labor, much less support a family. There was discussion about the formation of a miner's cooperative union to lease one of the abandoned smelters. The workers had insufficient capital so the proposal failed. Some mills were being removed to other districts; two or three were reduced to cinders by fire. As winter approached, the few remaining miners were faced with the choice of abandoning the district or trying once again to "wait it out" until spring. Word of new discoveries in the Utah mountains arrived at this juncture. Men stampeded in that direction in companies of a dozen or more, in twos or threes, or as solitary horsemen.[40]

Although the winter of 1870–71 found the White Pine District largely depopulated, the Eberhardt and Aurora Company continued its program of expansion. The International Mill was working successfully, and to keep it adequately supplied the company had purchased another noted mine, the Ward Beecher, that had produced some $200 ore. For this three hundred-foot claim adjoining the North Aurora Mine, Applegarth was given one thousand fully-paid shares in the company, valued at £ 10,000.[41] In spite of favorable reports from Nevada, the value of shares in the Eberhardt and Aurora Mining Company fluctuated severely during January, 1871, falling as much as £ 2/10s. in a few days. "Bear" raids continued throughout March and the £ 10 shares, once worth £ 25 on the market, were down to £ 15 or £ 15/10s. The South Aurora Company shares were simultaneously selling for a 10s. premium, or £ 5/10s.[42]

By April, 1871, George Attwood reported that the first three months profits at the International Mill were £ 15,000, and if this record was maintained, the directors would be able to declare a £ 2 dividend in July. There was a splendid body of ore in the Consolidated Aurora and Ward Beecher mines; the mill manager thought he had never seen ore of such richness since the early days of the Comstock. Sufficient ore was in sight, according to reports, to keep the new mill running for many months. Shareholders were called together in extraordinary meeting to sanction the issuance of 3,500 new shares, 1,000 of which the directors had promised for the Ward Beecher Mine and 2,500 more to be offered to the current shareholders at a £5 premium. The chairman reported the mine had "turned out something really wonderful" and within three months had more than paid for the Ward Beecher Mine. One large shareholder pointed out that future profits were to be much greater than the output of the Oasis Mill indicated. For example, the salaries of general manager, mine and mill superintendents, debited against the ore crushed by the small mill, would account for $6.11 for each ton whereas the large mill crushing one hundred tons of ore a day would account for only sixty-three cents. This was a very small part of the $20 a ton calculated for expenses when the original profits were projected. Meanwhile, Phillpotts was ill with mountain fever and his condition had been aggravated by mental anxieties and bodily exertions at Treasure Hill. Attwood had taken temporary charge of the company's affairs.[43]

The South Aurora Silver Mining Company was also prosperous. Ninety men were working at the mine and a larger ore house was being built to facilitate sorting the ore before it was sent to the Stanford Mill.[44] The company declared a quarterly dividend in February at the rate of twenty per cent per annum; a similar distribution was made in May. Some questioned the advisability of the last distribution because a severe storm in mid-February had made it impossible to deliver ore by wagons, and the mill had been shut down to await the completion of a tramway under construction by the Eberhardt Company. The mine, meanwhile, was being steadily worked and an ore reserve accumulated. The vendors agreed to "defer" the quarterly dividend payment on their shares in May because their contract with the English had provided a guarantee that the tramway would be available to carry this company's ore and that it would begin operation by February. Meanwhile, the directors had sold an additional twenty thousand £5 shares for £10, or a one hundred per cent premium. Applications for these shares were three times the number to be issued because the market quotation at that time was from £11 to £11/10s. a share.[45]

In this period of inflated values for White Pine mining company shares, several new British companies were promoted, most of which never processed any ore. The East Sheboygan Silver Mining Company, Limited, proposed to develop a property adjoining the Hidden Treasure Mine. The vendors agreed to accept paid-up shares in the company so that the entire capital raised in England could be used in development work.[46] This company issued 47,500 shares of £2 par value and when sufficient cash had been raised, elected to purchase the Silver Glance, southeast of the Eberhardt, rather than the claim first considered.[47] The

Great Western Silver Mining Company, Limited, had also been registered and an agent dispatched to Nevada to take possession of its property. The South Aurora Company was considering purchasing the
property of the Consolidated Chloride Flat Company, covering seven and
one-half acres adjoining its mine, for £10,000. The *Engineering and
Mining Journal* of New York reported,

> English capitalists are generally well pleased with their purchases
> of American mines. Nevada mines seem to be attracting great
> attention in the London mining market and the shares of some
> of the companies, which completed their purchase some time ago
> are worth four or five times the amount paid for them. The Eber-
> hardt and South Aurora mines have been very successful since they
> came into the hands of their present owners. The stock of the for-
> mer sells for 400 per cent, and the latter for over 200.[48]

The Eberhardt and Aurora Company, Limited, had undertaken
construction of a cable tramway to transport ore from the mines on
Treasure Hill to the International Mill in Eberhardt City. White Piners
learned during the summer of 1869 that such an invention had been patented in England by one Hodgson, and they realized that an elevated tramway would make it possible to deliver ore to the mills in the midst of winter
when snow blocked the wagon roads. First practical application of this
invention was in Leicestershire, England, where stone was conveyed
from a quarry to a railroad depot three miles distant. After noting the
advantages of a tramway over grading a road bed, building bridges and
filling gullies, either for wagons or rails, the *White Pine News* suggested,
"It is needless, in fact, to enumerate the advantages of this method,
since they are apparent to everyone. We think this subject one that is
deserving the attention of our large mine owners and hope some of them
will have the courage to introduce the wire tramway in White Pine."[49]

Vendors of the Eberhardt Mine convinced the London directors
that one of the best investments they could make would be in a tramway,
and a construction contract was signed with William Thairwall and
Company of Middlesex. Armed with extensive engineering sketches,
blueprints, and specifications, Thairwall came to Nevada to supervise
the building. Tons of wire and equipment were shipped from England,
around Cape Horn, to San Francisco and then hauled overland to eastern
Nevada.[50] As finally completed, the tramway of seven-eighths inch steel
cable ran for two and one-fourth miles between the mines and the mill.
The ore, placed in wooden buckets carrying about two hundred pounds
each, moved along the wire about three miles an hour. Near the mill
there was a sixteen-horsepower engine furnishing power for passing the
continuous rope around a drum returning the empty buckets to the

mine at an elevation two thousand feet above the International Mill. The cable passed over pulleys supported by some fifty frame structures, the highest of which was one hundred and thirty feet high and the lowest twenty feet. These posts were built of strong timbers, ten inches thick, braced and framed.[51] Telegraphic apparatus for signaling purposes was also installed. With the telegraphic equipment from London came instructions for its use:

> When in the morning the line is about to be started the man in charge of the driving terminus is to ring the bell at the other end of the line — this will at the same time ring the bell at his own end so that he can know at once if his signal has been given. When he has done this he must wait until he receives a signal by his bell being rung by the man at the other end — when he can start his line — on leaving the work at night he must fix his commutator down by screwing down the screw . . .[52]

Within a few months the company installed three lines of telegraph wire so that messages could be delivered along the tramway route. This conveyance was the largest cable tramway in the United States in the 1870's. The British had invested approximately $135,000. However, the company expended only one dollar a ton to transport ore in this fashion and the improvement was expected to pay well because it had previously cost three dollars a ton to deliver ore by road in the summer and five to seven dollars during the winter months.

As soon as the tramway was completed, technical difficulties developed. Temperature variations, resulting from the difference in elevations, caused an expansion and contraction of the wire so it sagged between some of the posts. This problem was unique in the history of the British manufacturer who designed and patented the apparatus. His representative Thairwall returned to Nevada to deal with the phenomenon and make the necessary alterations. By moving the tightening pulley half way down the hill rather than locating it on the crest, sixty feet of slack was removed from the 11,000 foot cable. As much as three hundred tons a day was now delivered to the mill.[53] It was soon found that the configuration of the terrain was such that the descent between some stations along the mountain side was so precipitous that ore carts crashed into the supporting posts along the way and spilled some of the precious cargo. Because of this loss, company directors resolved in July, 1871 to transport only the low-grade ore by tramway and cart the richer.[54]

Toward the end of 1871, shares in the Eberhardt and Aurora Mining Company began to fluctuate wildly. Rumors circulated that the rich ore had played out, and the London *Times* confirmed that only forty-dollar ore was left. As one commented, "the shareholders are in the dumps, in spite of the fact that Mr. Phillpotts says there is plenty of ore on the dumps." By the time the company got the tramway working properly, the boiler at the new mill burned out. The directors were immediately criticized for having built one sixty-stamp mill rather than two of thirty stamps; cost of reduction might have been greater but the risk of shutting down all operations much less. To reassure shareholders, a third quarterly dividend was declared in September on the basis of £1 per annum.[55] Investors in the South Aurora were likewise concerned because their shares had dropped 50 per cent in value during the four months prior to October, 1871, and were selling at £5/10s. At the second annual meeting, the directors admitted profits were meager, but some income had been secured by crushing the ores of the Eberhardt and Aurora Company at the Stanford Mill while the International Mill was being repaired.[56]

An extraordinary meeting of shareholders of the Eberhardt and Aurora Mining Company was called in November to authorize the issuance of fifty-six hundred £10 shares, primarily to purchase the White Pine Water Works, thereby guaranteeing sufficient water for the International Mill. The owners demanded £30,000 for the waterworks; the additional funds were to be used in sinking deep shafts hoping to discover the vein of rich ore found on the surface of the hill. These investments were deemed essential to protect the earlier expenditures at Treasure Hill. Several shareholders seized the opportunity to raise strenuous objections to the unbusinesslike tactics of the directors. Why, asked one investor, had the mining company paid for adjustments to the tramway? Wasn't the construction company liable to put it in running order? If the waterworks was such a sound investment, why wasn't a separate company registered to buy and operate it? A committee of investigation was called for, but the majority of the shareholders stuck by the Board.[57]

Seligman Brothers of London, who had promoted the sale of White Pine mines in Britain, also represented the owners of the White Pine Water Works. Possession of this water supply system would insure to the Eberhardt shareholders control of all other mills in the district, like the Manhattan, the Stanford, and the Dayton. In the end, a separate venture was formed, known as the White Pine Water Works Company, Limited, with a capitalization of £50,000 in £5 shares. There was an interlocking directorate with the Eberhardt and Aurora Mining Company. An initial payment of £10,000 was made, and a final settlement was due only upon proper conveyance of land title.[58] The deal was consummated in March, 1872.[59] Seligman Brothers, who worked closely with Leland Stanford in San Francisco and Nevada promotions, gained

Stanford's support in establishing the Anglo-California Bank of London and Paris the following year.[60]

Meanwhile, Manager Phillpotts was under fire because of his extensive land purchases, the price of which had, in some instances, been excessive. The title to others was questionable. One particular claim, valued at $1,200, had been purchased by Phillpotts for $8,000 on the assumption that its possession was vital to connect two shafts dug by the British corporation. Later developments proved the expenditure unnecessary. A serious legal complication developed over the Ward Beecher. Henry G. Blasdel, former governor of Nevada, claimed an interest in a portion of the area sold to the British by Edward Applegarth. The company's lawyer in Hamilton, Thomas Wren, who spent most of his time and legal talent trying to untangle the myriad of deeds and documents pertaining to the company's property on Treasure Hill, notified Phillpotts that he could only estimate the probable cost of a suit against Blasdel. Earlier litigation between the Eberhardt and Richmond mines had cost Drake, Applegarth, and their partners approximately one hundred thousand dollars, $30,000 of which was lawyer's fees. The South Aurora had spent $10,000 for attorneys and another $10,000 for expenses to quit the claim of the owners of the Autumn Mine to its holding. "For a mine like the 'Ward Beecher' at the present time I think ten thousand dollars would be a low estimate and it might cost fifty thousand dollars to conduct suit successfully," advised Wren.[61] Meanwhile, the Commissioner of the General Land Office rendered a decision permitting Blasdel to patent a claim for three hundred feet of the Ward Beecher deposit. The adverse claim of Phillpotts was rejected because it had been filed after the approval of the land survey and because there was insufficient evidence to establish fraud or lack of good faith on the part of Blasdel. The company announced that this patent decision of the Land Office in no way affected its suit before the Nevada Supreme Court to determine the legality of the deed made by Blasdel to Drake, Applegarth and others which they in turn had transferred to Phillpotts.[62] Before the court made a final decision the matter was compromised, but not until the British had expended thousands in court costs and legal fees.[63]

Phillpotts, apparently unaware that the British investors' patience was near exhaustion over his complicated land deals, wrote Melville Attwood in London asking whether he should advise the directors to buy some land adjoining their claims that one of his "upright and honest" friends was trying to market. Attwood told him the company had sufficient undeveloped mineral property, and that it would be unwise for them to lease or purchase more. "I regret to learn that an attempt is now being made in London to sell many worthless Nevada mining properties, particularly in the neighborhood of White Pine," he wrote. "I think it would be well to leave the matter to your friends."[64]

The British were justly concerned over their sizable investment in eastern Nevada. Between September, 1870 and October, 1871, total company expenditures had been $912,868. Company accounts revealed an outlay of $257,501 for the International Mill; $137,418 for tramway construction and running expenses; $114,014 for mine operation; and over $20,000 for legal expenses.[65] In December, "a working man" of Treasure Hill wrote the London *Mining Journal* that the rapid decline in mill profits was clear evidence that the surface deposits had played out. He predicted that by the beginning of 1872 there would not be one ton of valuable quartz left in the Ward Beecher Mine. The Eberhardt Mine, he thought, would not find a purchaser if it were offered for $500. There was always the possibility of valuable ore being found further down inside the hill. Unless this were the case, he predicted the English company's tramway would never pay for itself and the International Mill would have to shut down.[66] Another miner wrote:

> It was a matter of astonishment to the business men of Nevada who were well acquainted with White Pine, when the operations of the Eberhardt Company were commenced and carried out on such a grand scale, for this company took hold of their work after men of great mining experience, and with an abundance of capital at their disposal, had abandoned the district. All about were evidences of failure. The towns of Hamilton and Treasure City, from being overcrowded, became, at least as far as Treasure Hill is concerned, almost deserted.[67]

The bullion ledger of the company revealed that between September, 1870, and October, 1872, the silver bullion produced was worth $1,169,070. This was an impressive amount, but while the assay figures had estimated an average silver content of $50.31 a ton, only $38.84 worth of silver was obtained. Total expenses, including mining, hauling, water, milling, and salaries, were $32.07 a ton, leaving a very small margin for profit.[68] By April and May, 1872, the company produced only five bars of silver a week, valued between $7,000 and $7,500. These were shipped to Seligman Brothers in London by Wells, Fargo & Co.[69] Some attempt had been made to reduce expenses, but in the year following October, 1871, the company expended $622,419, more than two-thirds the amount of the previous year.[70] Much of the outlay went for the wages of forty-two men employed at the mill and the 137 workers on the Ward Beecher and North Aurora mines, including the superintendent and

Foremen in mine	3
Foremen on surface	2
Timekeeper (and clerk)	1
Blacksmiths	3
Helpers in shop	3
Engineers	2
Carpenters	2
Assorters	12
Top Hands (surface)	22
Car men	8
Wheelers and shovelers (underground)	18
Bucket landers	2
Miners	58 [71]

Shares in the Eberhardt and Aurora Mining Company had dropped as low as £3, £2 below par, in May. South Aurora shares, with £5 paid in were selling for £2/15s.[72] British investors in White Pine, already thoroughly alarmed, were in for a more serious shock. The International Mill burned to the ground on August 31, 1872, and the wire ropes of the tramway broke shortly after. In their days of disappointment, the two English companies fell to quarreling between themselves and the South Aurora Company refused to allow the Eberhardt Company to use its Stanford Mill because of a misunderstanding over the fee to be paid per ton.[73]

The third annual meeting of the Eberhardt and Aurora Mining Company, in October, 1872, was stormy. William Baxter, a large shareholder, spoke for those opposed to the Board, pointing out that when Phillpotts had first gone to Nevada he had insisted no working capital would be necessary because of the immediate return from the mine. Shareholders were asked to be content with a thirty per cent annual dividend, but instead they had received £1 paid partly from Ward Beecher Mine ore and partly from a premium on shares. Not one farthing had come from the Eberhardt or the North Aurora mines. Meanwhile, company capital had grown from £200,000 to £290,000, not counting outstanding debts. Phillpotts had based all his estimates on $40 a ton ore; the average value per ton was within a few cents of this. Why, then, was there no dividend? The answer seemed to be in the gross and improvident expenditures by an incompetent manager. Extensive experiments in changing from a dry to a wet crushing process and then back again at the Oasis and International mills had cost thousands of dollars. Water pipes had not been laid to the International Mill so it could be flooded in case fire broke out. Baxter proposed a resolution:

"That the shareholders have no confidence in the management either here or in Nevada, and the directors be and are hereby requested to place their offices at the disposal of the shareholders." Applegarth complained at being termed a "western adventurer" who had sold a worthless property, but agreed that bad management was the cause of no dividends. As a holder of two thousand shares, he wanted a larger voice in the management and insisted mining and milling costs could be cut 50 per cent. In bitterness, A. D. DePass, company chairman, remarked that the Board must indeed be fallen when in the midst of three to four hundred gentlemen not one had put in a single good word for the directors. He reminded the group that he was the largest shareholder and had never sold a share. Moreover, he was not a partisan, being neither a "Phillpottite or an Applegarthite." Applegarth had already demanded DePass' resignation and had threatened to wind-up the company if the final installment of £5,000 due him was not paid. This money had been paid and the chairman announced his hope that the company was free from the Nevada vendor's influence forever. He pleaded with shareholders to bring Phillpotts home for an explanation before dismissing him. It was agreed to poll the twelve hundred shareholders on the issue of a change in management.[74] Within a week, however, a compromise was reached whereby Baxter and a like-minded individual, W. T. Allen, were added to the Board. The poll was no longer necessary. Phillpotts was summoned to London.[75]

Following conversations with the re-organized Board, Phillpotts was dismissed with the understanding that his return expenses to White Pine would be paid by the company and that his salary would continue until March, 1873, when his duties would cease. At that time, Frank Drake, his successor, was ordered to pay him $5,000, and permit him to use the company house, a servant, and team for three additional months.[76] Phillpotts insisted he was due a bonus of £2,000 from the company because he had succeeded in getting the original cost of the mines reduced by £100,000, but company solicitors advised that his claim be rejected on both moral and legal grounds.[77] Phillpotts' replacement by Drake transferred management from British to American hands. The former manager soon found employment as resident director for another British mining enterprise, the Davenport Mining Company, Limited, in Utah, whose Board was presided over by his brother.[78] George Attwood also left the Nevada company in March, 1873, to manage the Emma Silver Mining Company, Limited, another heavily-capitalized Anglo-American company organized in London. His reputation for extravagance and wastefulness was enhanced during his association with that notorious and infamous enterprise, but a decade later he was in charge of The Montana Company, Limited, also British.[79]

The South Aurora Silver Mining Company had likewise made personnel changes. Management had been turned over to a large shareholder, a dentist by the name of Goodfellow, who had no mining experience. The ore body played out in 1872, and Dr. Goodfellow resorted to exploratory work with a diamond drill. Shareholders complained bitterly when no promising ore was found, insisted that the dentist-manager's salary of £2,000 a year was excessive, and objected to his independent decisions without consulting the London directors. Nevertheless, they agreed to finance another experiment on Chloride Flat.[80] The quarrel between the two British companies raged on with the Eberhardt Board agreeing to pay $12.50 for the reduction of ore at the Stanford Mill and the South Aurora organization demanding $15.00. The latter company finally agreed to a one dollar reduction per ton, and a joint meeting of the Boards was planned to work out an agreement profitable to both companies. It was about time. During September and October, the Stanford Mill had produced only £1,800 for its owners.[81] Vendors of this property had received £200,000 cash for their claim and had sold large blocs of the additional £100,000 worth of shares they had received whenever they could be marketed at a premium. At the close of 1872, the company passed a special resolution to wind up voluntarily and to approve a reconstruction known as the South Aurora Consolidated Company, Limited.[82]

The financial condition of these British mining companies reflected the general situation in the White Pine Mining District, according to the State Mineralogist, who reported:

> On the whole, mining operations have not proved so successful as in the former years. The production of bullion has fallen off greatly, and many inhabitants have removed to other sections of the State where business is more encouraging. The once populous towns of Hamilton and Treasure Hill [sic.] are now fast approaching abandonment. There still remains in White Pine District a small population of miners, who are engaged scraping together the scattered bunches of ore left in former workings of the mines, and in prospecting for other deposits. Considerable bodies of ore are still found in the mines of the English Company on Treasure Hill, but judging from the results of experience in mines of this vicinity, their supply will soon be exhausted.[83]

The Eberhardt and Aurora Mining Company, Limited, like many other British mining companies in the western states, made a significant contribution to the technology of the industry. The company's mill, reported to be the most elaborate and efficient in Nevada, had been designed in England. The elaborate tramway to conduct the ore from mine to mill, reported to be the longest in the United States, was planned by an English engineering firm utilizing the latest patents, and the manager came to Nevada to supervise its construction. However, while theory and design originated in England, its application was usually left to Americans. For example, essential machinery for the mill and some parts for the tramway were fabricated in San Francisco foundries, and mining and milling equipment was regularly purchased there. The history of this company in White Pine illustrates the assertion that legitimate British mining operations in the West employed the up-to-date knowledge of mining and milling experts. As long as their resources held out, the British spent with a lavish hand so that none could complain of niggardliness. When no one else had been willing to risk the capital, they purchased the defunct water works so the mills in the district could be kept running, to say nothing of the convenience maintained for the householders in the mining towns.

Like other English mining companies in the American West, the Eberhardt and Aurora Mining Company learned that foreign companies were always at a disadvantage in litigation before the courts. When an 1872 Congressional statute limited the patenting of mining claims to United States citizens or those who intended to become citizens, the directors of the company sought advice from one of the most able lawyers in the state relative to their rights as an alien company. The lawyer selected, a future United States Congressman, thought the titles purchased from United States citizens were safe, but he could give the British no assurance that they could check poachers if the violators were United States citizens.

Within the first two seasons there were indications that the size of the body of ore had been misrepresented by the American vendors. The investing British public suggested that once again they had been sold a worked-out American mine. The vendors insisted the ore was available but that the failure to locate it and to produce adequate profits for dividends was the result of bad management by men sent out from England. The directors were in an unenviable position of being caught in the cross-fire of complaint between vendors and stockholders. The experience of the South Aurora Silver Mining Company, Limited, was even more disheartening. Both the directors and shareholders agreed that they had been duped. The Stanford interests in California had unloaded the South Aurora mine on them just as it was about to be worked out, along with a mill for which they had no use. These circumstances had made legal reorganization imperative.

All British companies had to deal with the problem of long-distance management. Most companies felt they could not rely upon American managers. To their sorrow, the companies in White Pine learned that the Englishmen sent out to Nevada were incómpetent. The Eberhardt and Aurora Company finally agreed to substitute one of the American vendors as director of the mining and milling operation. With this change in personnel, all looked toward a new day.

BRITISH DETERMINATION AND ADVERSITY

The only systematic, legitimate mining operations carried on in White Pine during the winter of 1873–74 were those of the British owners of the Eberhardt and Aurora Mining Company. Persistent efforts were rewarded periodically by evidence that additional ore was present. The editor of the *Mining and Scientific Press* noted:

> While other companies have been scraping up the ground, scratching here and there for rich ore, this English corporation have been working intelligently and with a purpose. The history of the company mentioned could be repeated over and over again if the owners of property display the same amount of energy, or have a desire to make mining a legitimate business, instead of a means to inflate or depress stocks on the market.[84]

Frank Drake announced that 6,000 tons of ore taken from a new area of the North Aurora mine had been assayed at $56.78 a ton, an estimated value of $340,000. The owners of the Stanford Mill, in contracting to process the ore, guaranteed delivery of bullion to the amount of eighty-two and one-half per cent of the assay with any excess being retained. The bullion actually produced was worth $300,000, or 87 per cent of the assay, so the mill owners made a bonus of $17,000 in addition to the milling price of $13 a ton. In all, the South Aurora Silver Mining Company, owners of the Stanford Mill, had made $95,000, or one-third of the profits from the ore in the new discovery, so the Eberhardt Company resolved to build a new mill of thirty stamps.[85]

Workmen were employed during the autumn to clear away the debris of the old International Mill site in preparation for the new structure. By the end of 1873 the mill was in running order. Two months later the *News* editor accompanied by Drake, visited company mines on Treasure Hill and reported that a magnificent body of ore had been recently discovered. The mines had produced enough ore to pay for the new mill's construction as well as all the laborers at the mill and mine, so the company was still a self-sustaining operation. Funds for prospecting were desperately needed and if they were forthcoming from London, the newspaperman thought a second Eberhardt might be found that "will make our English cousins dance with joy."[86] The Nevada State Mineralogist confirmed the editor's views by observing, "A more healthy feeling is manifested than has been felt heretofore in regard to the future prospects of the mines on Treasure Hill." The state official admitted, "But for the energy displayed by the superintendent of this company, and his faith in the permanency of the mine on Treasure Hill, there would be little left now but empty houses in this once famous mining camp."[87] Drake managed to gain respect from everyone. As one Nevada pioneer expressed it — "Frank Drake can scarcely be called the 'father of the Eberhardt,' [but] he should at least go down in history as its fairy godfather, for due solely to his initiative and ability the camp was safe [saved] from being entirely deserted."[88] When Drake took over the management, ore deposits were definitely playing out, the mill had just burned, and total losses were at least $300,000.[89] In spite of encouraging developments, shares of the company nominally worth £235,000 had dropped to £99,875 by the end of 1873. This was about one-tenth of their highest selling price of £987,000. The British were so discouraged that they seriously considered stopping work.[90]

Edward Applegarth usually attended the company meetings in London, and late in August, 1874, he went to the annual session of shareholders determined to get funds for development work from investors who were equally determined not to advance another shilling. Applegarth had already tangled with the Board over the procedures to be used in raising more capital, so he appealed directly to shareholders for a seat on the directorate. The Nevadan claimed it was essential to get an English Board that would work in harmony with Captain Drake, who remained with the company not for the meager salary he was paid but in his determination to prove the value of the Treasure Hill property. Chairman DePass thought Applegarth should not be on the Board because he represented Captain Drake, and when differences arose between the directors and the management, there could not be a fair discussion. Drake himself was already a member of the Board, providing the necessary mining talent. When the showdown came, Applegarth had proxies representing five thousand shares, but supporters of the Board successfully moved for an adjournment of the meeting. In a final statement, the chairman expressed his pleasure in the delay because the Applegarth proxies "had not been obtained in that fair spirit that one Englishman should adopt to another." In appointing Drake as manager, the company had thought it had conceded everything Applegarth desired in "the nursing of his bantling," but nothing seemed to suffice.[91]

When the shareholders reconvened on September 6, 1874, the directors controlled 9,427 shares and Applegarth had approximately 3,300 a respectable number. Meanwhile, Drake, whose previous reports had all been sober reading, telegraphed that the mine was "looking splendid" and urged the election of Applegarth to the Board. T. G. Taylor, a heavy investor who supported the Americans, inquired why chairman DePass, when he admitted having no mining experience, insisted on keeping a practical miner like Applegarth off the directorate. The Nevadan rose at that point in the discussion to suggest that the re-election of the entire Board would mean that the company would have to look elsewhere for an American manager. The chairman shouted that the shareholders, as Englishmen, would not be coerced by Captain Drake however well and ably he had managed the property. In the end, the chairman and William Baxter were re-elected. When pressed by many of the assembled group to give Applegarth one of the two other vacancies, DePass announced that as an individual he could not sit with the Nevadan in conducting business.[92]

By November, DePass had resigned as chairman, and E. W. J. Ridsdale, the first chairman who had withdrawn from the Board three years previously because of his lack of confidence in Phillpotts, was returned to the chair. His first act was to request Applegarth's election to the Board, insisting that few American vendors had been as continuously interested in a property they had sold to the English as Drake and Applegarth. In five months, July through November, 1874, the company had made £20,000, and if weather permitted operations to continue through December, Drake expected to have paid all the company's debts. Meanwhile, shareholders agreed to issue £14,000 in debentures to raise capital for development.[93]

The other London company operating in White Pine was also having its troubles. For eighteen months after the voluntary liquidation of the South Aurora Silver Mining Company, Limited, and its reorganization as the South Aurora Consolidated Mining Company, Limited, the directors debated the advisability of abandoning the company property at Treasure Hill. Some wanted to invest in Canadian gold fields, but the majority hoped for an amalgamation with the Eberhardt and Aurora Company, thereby consolidating all the British interest in the district.[94] Maintenance costs for the idle mine and mill were £1,000 a year. When the directors approached Eberhardt officials about obtaining water from the White Pine Water Works to work the tailings at the Stanford Mill, the price asked for the water was deemed "fabulous." A controversy ensued over ownership of the pipes leading to the mill, the South Aurora Company claiming to have purchased them as an appurtenance of the mill, and the Eberhardt Company insisting they were a part of the White Pine Water Works.[95] Taylor, who held shares in both enterprises, continued his crusade for amalgamation at the June meetings of both companies. The chairman of the Eberhardt and Aurora Mining Company commented:

> With respect to an amalgamation with the South Aurora Company, if you amalgamate you do it to obtain some advantage — (hear hear) — and with respect to this amalgamation I can not see any possible advantage to this company. We have a good property, we have mines which are yielding well, and we have a mill which is working well. The South Aurora have a mill, and a mine which is at present worked out. I do not say that if they went on working they would not discover some body of ore, but if they, instead of frittering their money away in all parts of the globe, were to use their diamond drill in proving their own property it would be better for them, but it is absurd to come here when we are in a condition of prosperity, and ask us to get them out of the mire when they will not put their shoulder to the wheel.[96]

The company solicitor reminded the chairman that water running to the mill of the Eberhardt Company passed through four and a half miles of pipe that also supplied the Stanford Mill. Deeds of conveyance had been exchanged and examined by both London companies; the South Aurora had a prior sale contract, but conveyance of the waterworks to the Eberhardt Company more specifically included the transfer of the controversial pipelines. When Taylor brought up the amalgamation proposition again in December session, the Eberhardt chairman told him there was "not a ghost of a chance of this company ever buying the South Aurora property."[97]

For the first time since 1871, the Eberhardt and Aurora Mining Company showed a profit in the mining account for 1875. A final debt of £24,000 had been paid and an additional £9,000 made. Captain Drake had pushed excavations down 440 feet and uncovered another valuable pocket of ore. He had telegraphed, "Everything is very prosperous, the mill is running well, and the assay of June is about $60 to the ton. We have 800 tons on the mill dump, a fine body of ore in the mine which is equivalent to twelve months supply, and 300 tons on the mine dump." A remarkable change had come about in a single year.[98] Shareholders were frustrated, however, because the company had taken £425,000 worth of silver bullion from its mines, and had only paid one annual dividend. Nor could dividend payments be resumed until all the debentures were liquidated, so shareholders voted to call in each £10 debenture and exchange it for a £10 share in the company and an additional cash bonus of £2/10s. Approximately seventy debenture holders surrendered 1,327, £10 certificates under this arrangement.[99]

At this juncture Frank Drake proposed that the Eberhardt and Aurora Company should sink a tunnel into Treasure Hill in hopes of discovering the vein of rich chloride ore. London directors had concluded that neither American "experts" nor professors could be trusted, so John Wild, one of their number who had mining experience, was sent on a tour of inspection to Treasure Hill. While Wild was in Nevada, a rich bonanza was struck on Ward Beecher ground at the depth of 560 feet. Secrecy was immediately enjoined upon workmen of the Eberhardt and Aurora Company, but rumors continued to circulate that the size of the ore body continued to increase and would be immense. The Eberhardt and Aurora Mining Company began shafts elsewhere down to the six hundred foot level in an attempt to prove the contention that the vein might go downward from the surface of the hill. Many San Francisco companies were waiting for the English company to do the prospecting and then they hoped to reap the benefits.[100] The editor of the *Mining and Scientific Press* observed, "We are glad to see, after all their bad luck, that the company will probably be rewarded. We sincerely hope that all that is anticipated will be realized, as these discoveries go far toward restoring confidence in much abused White Pine."[101] "Of course this is welcome and joyful news to the citizens of Hamilton," wrote the *News,* "as this company at the present time is our main support, and everything that is beneficial to them is also a god-send to our people."[102] The San Francisco mining journal further informed its readers:

> At latest accounts the bonanza was still widening out and improving greatly. It is believed to be by far the largest body of ore ever discovered in White Pine county, and what is better still, it is of exceeding high grade. Taken right through it is believed that the whole deposit will average about $300 per ton. Considering the great depth of this find and its general trend into the mountains, there can hardly be a question that it is the most important development that has been made in Eastern Nevada for a long time.[103]

The county tax assessor also confirmed that the amount of ore taxed in 1875 was greater than any one of the three previous years. His annual figure of $863,839 compared favorably with $738,498 for 1872, $495,796 for 1873, and $598,891 for 1874. If the new strike proved to be a bonanza, the yield for 1876 was expected to double.[104]

Wild hastened back to London with the good news and confirmed the high regard that was held for Frank Drake in Nevada mining circles.[105] Upon this director's suggestion, Drake made plans to go to London to discuss his tunnel project with the Board and to attend the annual meeting. Oliver Drake, his brother, came from California to supervise mining operations during his three or four month's absence. The *News* paid another glowing tribute to the superintendent at the time of his departure, and observed, "We hope he will succeed in convincing his company of the necessity of running a tunnel through Treasure Hill, the economy of which must be apparent."[106] Citizens of Eberhardt staged a surprise party to "take affectionate leave" of the Captain, and the whole town turned out for the "eating, drinking, toasts and songs, and a deal of merriment" that continued into the morning hours.[107] Drake arrived in London toward the end of March, 1876.[108]

In April, shareholders of the Eberhardt and Aurora Mining Company were notified, in semi-annual meeting, that the half-year profits had been £11,000. All debenture indebtedness had been liquidated so that millstone no longer was around the neck of the company. Captain Drake asked for £30,000, primarily for machinery, to drive a tunnel into Treasure Hill in a final effort to determine whether a vein existed. T. G. Taylor strongly objected to the expenditure and pressed for a dividend. He argued that the company's £10 shares were worth only £7 on the market, but the declaration of a five-shilling dividend would return them to par value over night. Taylor insisted, "You know it is no use going on year after year, and apparently for ever, earning money, but only enough to pay management expenses, and never dividing any profit among the shareholders. (hear, hear)." As the demand for a dividend increased, the chairman interrupted, "I am certain Captain Drake will back me in this respect when I say that if you want to throw the property away you will take the money now in the shape of a dividend." In the end Captain Drake was called upon to speak:

> I believe you have the best property in Nevada, but we cannot open it unless we have money to do it with. I have worked there, gentlemen, pretty hard, but I have not had the means to work as I wanted to work. If a miner has the money with which to work he can save a good deal; it has cost me a great deal more to work the ore than it would have done if I had the means. ... If the shareholders see fit to divide the money it is not for me to say otherwise; I leave it to the directors to pay a dividend if the shareholders say so. I am not going to object, but I assure you I do not feel like going to work with my hands tied again. (Cheers.) I worked for three years — Mr. Applegarth and myself — to put the property on the market, believing we had a good property. ... I have not had the money to go to work and get drills, and so work with greater speed; but we have had to hammer the ore out, and this takes time. It is impossible to do more than a certain amount of work. You can only put so many men to work in this chamber; so it is with a mine. I firmly believe you have a good property if not I never would have stuck to it. I think so still, and

I believe that if the shareholders do what is for their interest, and drive this tunnel, you will have a very good property. (Cheers.) I do not know that there is anything more I can say.[109]

Shareholders agreed to go forward with the excavation. The South Aurora Consolidated Company was to bear a portion of the cost, approximately £5,000, because the tunnel would run underneath its claim as well as those of the larger company. The South Aurora Mine had lain dormant for many months and the company's £10 shares were worth ten shillings.[110]

While Drake was in London, the company was forced to give forty miners on Treasure Hill a temporary discharge. Ore dumps at the mine were running over and the weather would not permit removal of the ore to the mill.[111] When April arrived with no break in the weather, the local newspaper complained, "This has been the longest, coldest, and most severe winter the oldest White Pine inhabitant has ever seen here. There has scarcely been three consecutive days, for over three months, that have been fit for men to work in the open air."[112] Two weeks later, Converse, the local freighter, announced that the road between the mines and the mill was opened up so his teams could get through. It had taken the labor of thirty men shoveling snow for most of the winter and had cost him $2,000 to put the road in a passable condition.[113] Toward the end of April operations of the Eberhardt and Aurora Company were once again in full swing with one hundred men given employment.[114]

Drake returned to Treasure Hill in July looking hale and hearty and showing "unmistakeable signs of good treatment at the hands of his London friends." The tunnel was started by making an open cut in Mahogany Canyon, on the south side of Treasure Hill, preparatory to the arrival of a Burleigh drill and other necessary machinery. Plans called for a five thousand-foot penetration into the mountain to be completed within two years. The tunnel was to be seven feet high, nine wide, with double tracks laid on its floor. During the first thirty days a one hundred and three-foot penetration had been made by working two shifts of men, each eight hours a day. The local newspaper thought "making over three feet a day, through solid limestone is pretty good without any machinery. Push her along, boys, everybody here is anxiously awaiting the result of your labors."[115] However, work did not progress as rapidly as Drake had estimated and he departed for San Francisco in August to obtain more powerful drills.[116]

Meanwhile, Converse, the freighter, had located a townsite at the mouth of the Eberhardt and Aurora Company's tunnel and filed a claim in anticipation of selling town and business lots at that strategic location.

A New York company was also planning to tunnel into Treasure Hill. Enthusiasm for tunneling projects reached a fever pitch when a lump of ore taken from the workings of the Eberhardt Company, assayed at $4,110 a ton, was placed on display in a local hotel.[117] When the new drills arrived from San Francisco in October the tunnel was only two hundred feet deep, but it was now anticipated that not less than five to seven feet would be completed every twenty-four hours. The company had also decided to sink an incline shaft to meet the tunnel and this was down 842 feet. Periodically valuable pockets of ore, though small, were encountered. More men were employed in the British enterprise than at any time in the past three years. Week by week progress reports were made as the tunnel reached 500 feet, then 600, and after the first of the year, 800 feet. Tension mounted as each "strike" played out.[118]

Shareholders in the Eberhardt and Aurora Mining Company at last received a dividend of five shillings in July, 1876, and shares immediately rose to £10 par.[119] Another dividend of three shillings was declared in January. When shareholders urged a larger distribution the directors reminded them that the dividend had been approved over the objections of Captain Drake. The superintendent feared that the pocket of valuable ore discovered and worked during 1875–76 was about to "pinch out." Moreover, the price of silver had dropped enough during the year to reduce the earnings on every share by ten shillings. Expenditures on the tunnel had been so great that the mine and mill had been shut down.[120]

Drake telegraphed London that the mine was not looking too well. The directors immediately wired back, "You telegraphed mine not looking so well, does this refer to the general aspect of the mine or to new ore body first level, please explain." Drake replied, "Refers to new ore body only, incline looking more favorable."[121] As the tunnel reached one thousand feet, the ventilation became so bad that time had to be taken to construct an air flume so the men could continue work.[122] When time came to start the mill in the spring of 1877, there was not enough ore accumulated at the mill to justify resumption of work. The tunnel had been driven 1,140 feet and Drake wanted to burrow 800 feet farther to get under all the major claims on the hillcrest. The few silver bars that had been sent to England had brought less on the market than anticipated, for silver worth 59d. an ounce at the beginning of 1877 had dropped to 47d. nine months later. The discouraged were reminded that the company had taken £629,000 sterling from the mine since 1870, but shareholders thought this little consolation when almost £600,000, or $3 million had gone back to Nevada for development work.[123]

The results of 1877 had been very discouraging for the English. Annual losses amounted to £24,190, and the reserves built up during 1875–76 were reduced to £7,000. Not more than 1,500 tons of ore remained to be processed and the maximum income from it would be £6,000. For two years Drake had been sending weekly telegrams to London about ore prospects in the tunnel and incline. When favorable reports "leaked out" to the press, shares would jump to £12 only to fall back shortly to £3. Market speculation was rampant on such occasions, and the directors were periodically forced by public opinion to deny responsibility for releasing information. By the spring of 1878, the tunnel was 2,856 feet deep, built at a cost of approximately £15,000 — almost £6 a foot.[124]

Captain Drake was now convinced that the mine's resources were exhausted and he cabled London that he did not feel justified in further development work. He was summoned to England in October, 1878, and the directors proposed that exploration work continue. Drake agreed provided the company would send an English engineer to advise and to share responsibility in the endeavor. Funds were again exhausted so the directors proposed to issue a maximum of £20,000 of debentures, drawing 10 per cent interest, redeemable in two years. Purchasers were to have the privilege of converting each £10 debenture into a share at any time. [125] The determination of shareholders was reflected in the remarks of one:

> Now, to suppose that this ore does not go down is to suppose an absurdity, and I have not the slightest doubt that if you follow it it will put you into rich ore, and if that be so your shares will go up to £10. If you will not subscribe a small amount of money and throw up the sponge, it is a very foolish business. I intend to take some of the debentures, and should it happen that the worst comes to the worst, and that ore has not been come into, I consider it would be a very fortunate thing for the debenture-holders, for I would be very glad, if I could afford it, to have the whole of the property, lock, stock, and barrel, for it would sell for double or treble the amount in San Francisco. I think Capt. Drake should know that I intend to become a debenture-holder, and I have been on the property and seen it.[126]

Meanwhile, expenses had been reduced to the lowest ebb possible. The directors worked without compensation in 1878, and the officials in Nevada had voluntarily taken a 20 per cent cut in salary. Expenses at the mine — salaries, insurance, and taxes — were only a third of what they had been two years previously.

The South Aurora Consolidated Mining Company finally gave up in 1878. The directors had paid the Eberhardt Company £1,000 as an initial installment on their share of the tunnel expense when that project was begun. When the tunnel reached a point beneath the South Aurora claim a second £1,000 had been due, but the company defaulted. The Eberhardt Company was prepared to attach the Stanford Mill for the remaining £4,000 due on the contract, if ore was ever found again at Treasure Hill. Although the South Aurora Company abandoned eastern Nevada, it continued explorations on a world-wide basis as the Consolidated Mining Company, Limited.[127] Applegarth was still a director in this enterprise capitalized at £100,000 in £1 shares.[128]

During 1879, the Eberhardt and Aurora Mining Company received £16,000 in silver bars from the Nevada mines; but there had been a considerable loss on the year's work. All the money raised by debentures in 1878 was now gone. The incline was down 942 feet but to form a juncture with the tunnel under the Eberhardt claim would necessitate going under the South Aurora claim. It would be pointless to come to a standstill, yet the company could not drive the drift through another's ground without paying a fee. Right of transit granted several years earlier had applied only to the tunnel. The directors told the shareholders that they had come to the end of the road. All funds placed at their disposal had been devoted to the search for the source of wealth found so abundantly on the crest of the hill. No one could see the interior of the mountain. They had utilized the best mining experience and advice obtainable, yet the results had been a bitter disappointment. The company now appeared to have a worked-out mine on which at least £400,000 had been expended, they had a mill that cost about £50,000, and the White Pine Water Works for which they had paid £30,000. Another £60,000 had been lost in the burning of the first mill that was constructed. The British had practically nothing to show for an investment of over $3 million.[129]

When the State Mineralogist visited Treasure Hill again he summarized the situation:

> The work of exploration for new bodies of ore in the mines of Treasure Hill has been very energetically prosecuted during the past two years [1877–78], but work has at last been suspended. The supply of ore from the old chambers had been exhausted, the product having long since fallen short of the expense of extraction and the work of exploration. No company in the State has been more persistent in its efforts to develop paying bodies of ore than the Eberhardt and Aurora Mining Companies, whose property is situated in the district.... Unless work be resumed by the Eberhardt and Aurora Companies, the abandonment of the mines on Treasure Hill will be soon complete.[130]

In February, 1880, Drake wrote his partner Applegarth in London, that a rich pocket of ore had been hit in the tunnel 4,875 feet from the entrance. On receipt of the news, shares in the Eberhardt venture jumped 100 per cent on the London market.[131] The tunnel was pushed directly through the vein matter and no attempt made to prospect it until the Eberhardt and Aurora Company and the Consolidated Mining Company could come to terms.[132] During the summer Drake explored the side drift at the joint expense of both companies, but his semiannual report of September admitted to the British sponsors that he had "nothing new or startling to record."[133] Commenting on the frank statement of the manager, the London *Mining World* opined:

> A good man struggling with difficulties is said to be a spectacle for the gods. This, of course, refers to a good man above ground, but Capt. Drake, burrowing to the deep bosom of the hill, must be . . . a touching and a moving exhibition of human perseverance under most adverse circumstances. The substance of the report may be stated in a word. Captain Drake has done his best, but candidly owns that as yet he has discovered little or nothing worth having. Now, though the hearts of the shareholders may have become sick with hope deferred, they would do well to continue to their superintendent that confidence which they have reposed in him in the past, with the full assurance that no change in the management abroad or the directorate at home would at all improve the position of affairs.[134]

At the end of October, the tunnel had been completed to its full length after four years of steady work and an expenditure of £50,000. Every pocket of ore that had been found proved small and had soon "pinched out." At the annual meeting in December the shareholders voted to "wind-up" and incorporate a new company known as the Eberhardt Company, Limited, with a capital of £210,000 in £1 shares. This decision was necessary because debenture holders were threatening legal action and the sheriff in White Pine County was ready to foreclose because of delinquent taxes.[135] The chairman telegraphed Drake, who was prepared to shut down, that the reorganization had been unanimously approved and money forwarded for him "to drive along and explore." Right of transit had been secured from the Consolidated Mining Company, and with an additional $200,000, the determined British hoped to connect the incline with the tunnel.[136]

Perennial rumors about the discovery of a rich body of ore in the Eberhardt and Aurora Company's tunnel — none of them based on fact — were publicized throughout 1881 in the Nevada press.[137] The Eberhardt Company continued to operate at a loss and Captain Drake was notified in July, 1882, to stop reducing low-grade ore and concentrate his time and money on uncovering some that would pay. Shareholders were called upon to pay an assessment of two shillings a share, thereby raising £5,000 to keep the endeavor going. Once again the chairman paid tribute to the indomitable perseverance of Drake, who out of loyalty to the company, had refused offers double his present salary.[138]

In 1882, Drake was called upon by the London directors to go to Montana and examine the potentialities of the Drum Lummon Mine. Within twenty-four hours after the arrival of his enthusiastic report in London, English interests agreed to purchase the property for $1,500,000 and cabled $500,000 to New York as an initial payment. This property proved to be the most successful mining investment of the British in North America. Within three years, stock of the Montana Company, Limited, was worth $50 million and Lombard Street had a corner on the world's supply of silver.[139]

Meanwhile, at Treasure Hill, Drake had taken all the rock drills out of the mine because the company could no longer afford to operate them. The Board of Directors had broken up. Applegarth had returned to the United States on private business. Baxter was financially ruined by commercial misfortunes, and Wild was physically exhausted.[140] Early in 1884, Drake and Applegarth wrote the company that it was essential to reintroduce the air drills because hand operations were too slow and unproductive. A minimum of £15,000 was needed. Opinion was sharply divided among shareholders on the Board's proposal to issue £20,000 in debentures for this purpose. The company had been reorganized only three years earlier to get rid of such an obligation. After prolonged debate, the majority agreed to support the Board. Only £3,300 was raised although subscribers were guaranteed their interest and the right at any time to convert the debentures into stock. Drake proposed that his salary be cut from £2,000 to £1,000. The English sent him £5,000, all they could raise, for development work rather than to close down the company.[141]

The Eberhardt Company finally gave up in 1885. Chairman Ridsdale resigned, following his other colleagues into retirement, claiming that rheumatism in the knee prohibited his attending the annual meeting. Professor Thomas Price of San Francisco had been employed to examine the company property and determine if further expenditure of funds was warranted. He estimated that $120,000 would be necessary to explore the unprospected ground, at a rate of $7,000 a month; to overhaul the mill would cost $25,000, and ore bins costing $10,000 would have to be erected at the mine. Although he was of the opinion that the company was "fully justified in making a further expenditure," the shareholders had no enthusiasm left.[142] The British had underwritten the mining interest in White Pine County for at least $5 million. Now they had nothing to show for their efforts, although some statisticians claimed that the mines of Treasure Hill alone had produced $22 million in silver by 1887.

Of all British investments in western America, mining was the biggest gamble, and stockmarket manipulators, whether bulls or bears, enjoyed their sport both in London and in San Francisco. Speculation in shares of the Comstock, Reese River, and White Pine mines of Nevada was probably more rampant than elsewhere. When the fortunes of British companies at White Pine began to decline, factionalism had developed in the Board of Directors, American vendors and their supporters clashed with the majority of British shareholders, and annual meetings were stormy affairs.

All the problems of management, litigation, speculation in shares, and bad luck could have been overcome if the silver ore in Treasure Hill had not played out. British companies were forced to admit that the fabulously rich silver ore scooped from the surface had been exhausted and the long-sought vein had not been uncovered. With dogged determination they tunneled into the midst of the hill to find it, gambled a fortune and lost. Unlike the Comstock, no "bonanza" was unearthed in the bowels of the earth to revitalize the fortunes of the district. The British had searched on until there was no hope. Although the shareholders had lost a fortune, the company had sustained the mining interests of eastern Nevada for over a decade.

ABANDONMENT

Mining developments in eastern Nevada, particularly in the White Pine District, were inseparably intertwined with those of western Nevada. The discovery at Treasure Hill had occurred at the end of a period of recession around Virginia City. Prosperity began its gradual return to the Washoe mines during 1868, the first season of the rush to White Pine. For the next five years the yield of western Nevada increased, largely as a result of the workings of the Crown Point mine. The annual production of $13,500,000 for 1872 appeared meager however, when the "great bonanza" was found the next year. Several years earlier, two Irishmen with California mining experience, James G. Fair and John W. MacKay, had formed a partnership with two San Francisco liquor dealers, James C. Flood and William S. O'Brien, to acquire the Consolidated Virginia and adjoining California mine. By sinking an exceptionally deep shaft they uncovered the incomparable Comstock Lode. By 1878, the

Consolidated Virginia had produced $60 million and between 1875–78, the California added another $43 million. The "Silver Kings" used their wealth to dethrone William Ralston and William Sharon as financial rulers of the western Nevada mines. The Bank of California had collapsed in 1875, partially due to speculation in the shares of the Comstock mines.

In addition to developments in western Nevada, the Reese River mines continued productive during the years of the White Pine excitement. As Austin declined, the Eureka District took its place. Attempts to refine the silver-lead ore of Ruby Hill had met with repeated failure between 1865 and 1868 and most miners there had abandoned the camp to go to White Pine. However, in 1869 two returning White Piners assured Eureka success when they erected a blast furnace that overcame the technical problems in recovering the silver. In 1869, bullion production was less than $100,000 but each successive season the amount was increased and in 1875 alone the Eureka District produced $6,100,000. In less than seven years $20 million in bullion was attained. The milling center at Eureka had a population of five thousand in 1880 and the mining camp at Ruby Hill had over two thousand.

New mining districts sprang up all over eastern Nevada between 1870 and 1876. Pioche was built in 1869 as a center for mills and furnaces to service the mines of a half-dozen districts in the mountains between Pahranagat Valley and Meadow Valley Wash. In the winter of 1870–71 it was the most active and important mining town in southeastern Nevada and three years later could boast of a population of six thousand. In White Pine County a new and successful mining district was organized at Cherry Creek to the northeast of Treasure Hill on the slopes of the Egan range during 1872. The same year discoveries were made toward the southeast, and four years later the town of Ward was built with a population of fifteen hundred.

How could the White Pine Mining District survive such competition throughout the state?

———— • ◄ • ————

"White Pine at its birth. . . was a prodigy," wrote the editor of the *News* in 1874. "Electrified by the prospect of interminable wealth, multitudes took their line of march for the new Mecca, and so great was the hegira, that, . . . some ten or twelve thousand men established themselves in huts and caves, nine thousand feet above the sea."[143] A revulsion followed the excess at Treasure Hill. The speculative operator withdrew and, to account for his decision, began disparaging the district. The prospector, baffled by the region's geology, also took up his belongings and left, joining the criminations by loudly proclaiming that true mineral veins did not exist. The excess of mills over available chloride ores, the failure of furnaces because of inefficient construction, and the lack of appreciation for lower grade ores, conspired to depopulate the district. At least thirteen furnaces, started up like mushrooms, had, like mushrooms, vanished.[144]

The deteriorating mining picture had sounded the death knell to municipal government in the district. Real estate values had topppled so much after the boom seasons of 1868 and 1869 that the tax revenue of Hamilton, assessed in July, 1870, was less than half the anticipated amount. The entire income would not begin to run the wheels of local government; the pressure of debt from previous lavish administrations had become so burdensome that there was little left for current expenses. Hamilton was not alone in her misery, for Shermantown's finances were equally bad, and Treasure City's worse. Disincorporation did not appear possible.

> If we were to close up municipal shop today, what assets have we to meet the outstanding indebtedness of $30,000? We have the City Hall, an extensive fire machinery, an engine house, and a lot of street assessment dues not worth the paper they are written on. Could we realize $10,000 on the entire caboodle? Possibly. Who then would guarantee the balance? The question has never been brought to a practical test; but it is a principle of law that the individuals of communities are liable for public debts, the same as partnerships. There may possibly be a chance to test this requisite condition of law in White Pine. . . . How would it do for the three towns — which, after all, are the county — to throw their liabilities and assets into a general pot, and to require the next legislature to fund the entire amount as a county debt; and thereafter let the Board of Commissioners manage the entire affairs? We throw this out as a feeler. It may bring other and better suggestions to the surface.[145]

The 1871 session of the state legislature passed a series of acts to relieve the financial embarrassment of the local governmental units in eastern Nevada. The Board of Trustees of Treasure City was ordered to issue no more warrants. Three-fourths of the town's income was to be placed in a "redemption fund" to extinguish its debts; the remaining one-fourth could be used for current expenses. A scheme was legalized whereby each time $500 was accumulated in the redemption fund, the city fathers would advertise for sealed agreements to purchase outstanding warrants at a specified figure. Those who would settle with the town for the smallest percentage on the dollar would be paid off first.[146] A similar arrangement was made for White Pine County providing that 30 per cent of the current income would be used to pay indebtedness and that bids would be received each time $1,000 had accumulated in the treasury.[147] Hamilton got a similar financial boost by state legislation.[148] In 1873, the state of Nevada also released its claim upon any money that might be regained from the absconding county treasurer, Lewis Cook, or his insolvent bondsmen by agreeing that such funds should be deposited in the county treasury.[149]

By summer, 1873, it was apparent that White Pine would not have its seasonal boom. To the traveler, Shermantown presented "a spectacle of gloomy desolation equalled only by Goldsmith's description of the 'deserted village.'" On the road between that camp and Hamilton were the abandoned remains of the Little Giant, Kohler, and Metropolitan mills. More ruined structures dotted the hillsides below Treasure City. Only the mines of the English were being systematically worked. The local newspaper admitted "we have had a hard and long siege of dullness, and it will take some time to recover entirely from the effects of the black eye received in days gone by," but insisted, "taken altogether we can see nothing discouraging in the appearance of the White Pine District as it stands today."[150] However, the great Hamilton fire of June, 1873, destroyed the town as a business center. Only five hundred people remained. Treasure City had three hundred, Shermantown not over twenty-five, and the English company town of Eberhardt, approximately one hundred.

The *White Pine News* admitted that times had never been so dull as in the winter of 1873; along the narrow Main Street of Hamilton, once thronged with people, "one could now fire a hundred-pounder, loaded with canister, its entire length and not hurt a man." Non-resident capitalists in San Francisco were held responsible for this situation by the failure to work their properties. The Edgar Mine owned by the Ward Beecher Consolidated Company was a case in point. The last ore the company had run at the Manhattan Mill had been worth from $30 to $50 a ton, yet the company had ceased milling. Instead, shareholders were asked to pay a $7 assessment on shares not even quoted on the market. The *News* insisted this "system of freeze-out, now so common, must cease."[151]

When the Eberhardt and Aurora Company uncovered ore of sufficient value to build a new mill at the close of the year, the gloom quickly disappeared and the editor's usual optimism caused him to comment:

> It is only a question of time, short at that, when our camp, now so dull and deserted looking, will "loom up" as it deserves, and take its place among the very best mining sections of the known

world; and to those who have clung to White Pine through all the years of apparent decay will be due the praise for persistency and faith, and their reward will be certain.[152]

Bullion statistics did prove encouraging. Not a pound of silver had been shipped for ten months following the International Mill fire, but from June, 1873, to March, 1874, bullion worth $160,000 had been forwarded by express through the White Pine County Bank. The Eberhardt and Aurora Company alone had sent $80,000 to London in the two months since its new mill had started operation.[153] A. J. Brown, a mining engineer, arrived in Hamilton in March, 1874, hunting statistical information for Rossiter W. Raymond's governmental report on the mineral resources of the western United States. His calculations indicated that the White Pine District had produced 186,048 tons of ore up to January 1, 1874, for a gross yield of $8,767,484, or an average of $47.07 a ton. The local newspaper editor hailed this record and seized the occasion to express again his conviction that the White Pine District would regain her former glory.

> We regard this county to be just on the turn in her history, and confidently believe that before many months our bullion product will again astonish the mining world. We have men left here who have struggled on through years of adversity and borne up under the disappointment of hope deferred, brought on by the acts of unscrupulous speculators and their damning influences, who still *know* that success awaits them and are content to stay until the future shall work out their salvation.[154]

"Bull Run will be forgotten in the march to the sea," he later exclaimed.[155] The Nevada State Mineralogist also admitted that there was a slight increase in the amount of bullion produced in the district during 1874 over that in 1873 although his annual totals of $502,516 and $494,596 were somewhat more conservative than the typical statement.[156]

The town of Hamilton was finally disincorporated in 1875. As had been suggested five years earlier by the press, the Board of County Commissioners for White Pine County was named the Board of Trustees for the town of Hamilton. The legislature established the Hamilton "debt fund" providing for a special annual tax of fifty cents on each $100 evaluation of property within the former corporate limits of the city. This fund was to exist in perpetuity until all debts had been paid.[157] The census enumerated a total population in the county of 2,557, indicating a loss of 4,625 persons in the last five years. The Census Marshal also revealed that the dwellings listed for tax purposes were 747, but many of these had been abandoned.[158] The county had operated at a loss for 1875; the auditor reported income had been $52,166 and expenditures $53,024. Major outlays had been made in transcribing the records of a portion of Nye County transferred to White Pine and drawing the new boundary between White Pine County and Lincoln and Nye counties. If these extraordinary expenditures had not been necessary, he thought the county might have started paying on its debts.[159]

The heaviest concentration of population in the county was now at Cherry Creek and Ward. There were only 175 registered voters in Hamilton in 1876, 81 on Treasure Hill and 78 in Eberhardt.[160] The annual income of the county was a few hundred dollars more than the annual expenditures, but this did not go far in extinguishing a debt of $132,812. Nor did much relief appear possible. The county assessor had levied taxes on only sixteen mines during the year totaling $2,214. Of this amount, the Eberhardt had paid $1,821. Half the mines were assessed at not more than $8 each.[161]

The winter season was always a time of depression, even in the boom period. Each year when the snow began to fall more mines closed down for the cold period when mining and hauling of ore were exceptionally difficult. In 1875–76 the San Francisco owners of the Hidden Treasure, Mammoth and other mines had, as usual, ordered all work stopped for the winter months. The *News* advised the many miners thrown out of employment that they should form small companies and lease some of the good properties lying idle and work them through the cold season. This would be more advisable than spending their former earnings in hunting work elsewhere.[162]

The British company always kept a small force working through the winter and announced its determination to do so again in 1876–77. However, Drake finally decided to let the last thirty-five miners go and, as earlier noted, to concentrate on the excavation of the incline to form a junction with the tunnel through the middle of Treasure Hill. Many of the discharged men, ignoring the suggestions of the *News* editor the previous year, left for Eureka, Austin, and Cherry Creek trying to find employment.[163]

A. Skillman and Fred Elliott, who had purchased the *White Pine News* in 1873 from Forbes and his associates, reduced the paper in size, and changed its political allegiance to the Democratic party. From this time forward the *News* struggled to exist. In 1875, Elliott withdrew from the firm, but Skillman continued to publish the paper as a tri-weekly and then as a weekly until November, 1878, when regular publication ceased. After that date his interest was concentrated in the Eureka *Sentinal*. Repeated efforts to revive the systematic publication of the newspaper in the next two years failed.[164]

Between 1859–80 the bullion extracted from all Nevada mines had an estimated value of $306 million and the dividends declared were $118 million. Although the mines at Treasure Hill made a noteworthy contribution to this bullion production, the silver had been used to promote the mining industry rather than provide wealth for discoverers or investors. By 1878, the great mines of the Comstock had begun to decline. Activity around Pioche had slowed down as early as 1875 and the next year the town was largely abandoned. Cherry Creek, like Hamilton, had known prosperity for only two years and by 1875 large segments of its population moved on. Ward lost two-thirds of its population in 1877. Thus, most of the mining districts of Nevada — in the west, in the center, and in the east — joined the White Pine District in its demise. The great decade of Nevada mining, the 1870's, had come to a close.

————•—•—

At the time of the 1880 census only 117 registered voters remained in Hamilton, forty-nine at Eberhardt and fourteen at Treasure City, indicating a further decline of the White Pine District. The mining communities of eastern Nevada still attracted a grand army of bachelors. The population also remained heterogeneous. Italy had the lead among those of foreign birth, with Great Britain, Germany, and Canada generously represented. Mexicans and Chileans worked as packers or divided the community's business with the Italians. Most of the Irish miners had moved away. In the entire county there were only four lawyers, four doctors, and not a preacher of any sect.[165]

An estimated 250 miners still refused to leave the White Pine District in 1881. The principal mines thought to have some potential besides the Eberhardt, North Aurora, Ward Beecher Consolidated, and the South Aurora — all British owned — were the Stanford, Central, Hidden Treasure, Imperial, Mobile, Trench, and Jennie A. Work on the last of these Treasure Hill properties had been given up in 1878, and Treasure City was abandoned for all practical purposes, containing only one family home and a few single miner's cabins. A solitary family comprised the entire population of Shermantown, so the limited business activity was confined to Hamilton and Eberhardt City. The British kept the camp at Eberhardt alive. That settlement had a general store, a blacksmith shop, a carpenter's shop, a post-office, and an active temperance organization. The mail was brought in tri-weekly from Hamilton by stage. The thirty-seven children in the district attended school in Hamilton.[166]

Treasure Hill residents never solved the problem of isolation and inadequate transportation. The nearest railroad station was Eureka, southern terminus of the Eureka and Palisade Railroad, forty-three miles northwest. In 1876, a rail line known as the Eureka and Colorado Railroad was projected from Eureka to Pioche, on the headwaters of the Colorado River, but it was not located to run through Hamilton. Although some Treasure Hill citizens hoped a spur line would be built, traffic did not warrant the outlay of funds. The stage picked up mail at the railhead three times a week. All freight shipped to or from Eureka cost $20 a ton.[167] A Hamilton correspondent writing to the Eureka *Leader* illustrated the grim determination of the remaining residents to be loyal to the end.

> Our town looks cheerful, no one is sick, a poor place for doctors; all from the infant to the aged are in blooming health. Our school is in a flourishing condition, with some twenty-five pupils in attendance. . . . Our merchants look smiling, demonstrating that they are willing to trade anything from a jack-knife to a side of bacon.
>
> The citizens of H. and all the surrounding country are laboring under great inconveniences since our mails were reduced to a tri-weekly. Under the old regime we received our mails from the Bay City in from five to six days, now it takes ten.
>
> In conclusion I would say that the citizens of Hamilton do not despair. We have lived under dark and dismal clouds of adversity; all seemed lost. We now think and believe that the day is not far distant when old Hamilton will raise from her ashes.[168]

The *White Pine News* sporadically published a four-page weekly. Each time the newspaper announced an issue as the last one, friends came forward to help out financially. The editor paid them public tribute.

> We are glad to say that the *News* has prospered beyond our most sanguine expectations, and that our heads are above water, and we propose to keep them there. We are out of debt, have money in the treasury, and more in sight, hence we can stand alone, but, for all that, we shall always keep a warm place in our hearts for those who generously backed us in the venture without hope of reward.[169]

At the close of 1880, the editor announced that the *News* was moving its publication headquarters to Cherry Creek. "We are sorry that circumstances force us to move our paper . . . but we can see no way in which to make it self-sustaining here," for the bright hopes that "encircled White Pine Mining District have vanished in thin air."[170]

Newspapers throughout Nevada reported that an effort would be made to remove the county seat from Hamilton to Cherry Creek in

1881, but the *News* insisted that no such plan was being promoted by the citizens of Cherry Creek.

> They have far more important and lucrative business to attend to at present. Perhaps after a while, when our population increases to seven-eighths of that of the entire county, and our tax roll to nine-tenths of the assessment roll, we may grow tired of making semi-annual pilgrimages in mid-winter to the present snow-bound capital of the county. For the present we hold our souls in patience.[171]

The Post Office Department had been threatening a general reduction in mail service throughout Nevada since 1880 and early in January, 1882, Hamilton was reduced to a weekly connection with Cherry Creek and Eureka.[172] Agitation to remove the county seat was renewed and one resident wrote the *News*:

> Now that we learn you Cherry Creekers have inaugurated steps toward the removal of the county seat, even the Court-house clique appear ready to accept the inevitable. Some few, of course, feel in bad humor, and denounce the News for advocating what appears to them such a monstrous measure as making the officials go where the people are, instead of making the people come to them. . . . Of course, we all know the county seat should be removed to some more central point, and it is not likely that any effort will be made to thwart the will of the people in this respect.[173]

Late in March the snow at Hamilton was still eight feet deep and mail had to be dragged in over the snow on an express box, the driver riding one horse and leading another. "Truly a nice place for the Court to meet in the week after next," commented the *News*.[174] At an election in April, 206 of the 327 voting taxpayers favored removal of the county seat from Hamilton.[175] The county commissioners insisted that a petition of three-fifths of the voters, "each voter being a taxpayer," must ask for a change of the county seat and agree upon a new location. Various towns aspiring to the designation fell to quarreling among themselves and the movement temporarily failed.[176] In the spring the Eureka *Leader* announced that another strike had been made in the Eberhardt and that miners were talking about returning there. The *News* observed:

> It is a great pity that some of the rich bonanzas the Eureka papers are every once in a while bringing to light in Treasure Hill do not pan out as represented. We are getting tired of these periodical fits of romancing our Eureka exchanges engage in. . . . The showing is certainly not of a nature to induce old White Piners to rush back there. Now, honest Indian, wasn't the report circulated to influence county seat matters?[177]

No one was working in the Treasure Hill mines in the summer of 1882 except twenty-five employees of the British company cutting drifts from the tunnel.[178]

Hamilton was now waging a hard, uphill fight for survival. In 1885, a disastrous fire destroyed the county buildings and most of the town. The county seat was then moved to Ely.[179] Although the days of glory were gone, many of the community's outstanding personalities lingered on. In fact, Hamilton weathered the dull days following the boom primarily because of the loyalty and unflagging belief in her future by many old-timers. The stage between Hamilton and Pioche was operated by Gilmer and Salisbury. After completion of the Eureka and Palisade Railroad in 1876, these men also established a stage line between the railroad and Hamilton. The town's well-known merchants, James Mathewson and P. Evarts operated Evarts and Company until 1876. Mathewson eventually became one of Nevada's most successful businessmen, and Evarts superintended the Eureka and Palisade Railroad. A well-known character in town was "Colonel" Joseph Grandelynr, an authority on metallurgy and lexicography, who always met the stage to tell newcomers about the advantages of the district. When the economic decline set in, the "colonel" was employed as a night watchman by the San Francisco owners of a mine and mill who continued to pay county taxes and insurance. The story is told that when the great fire broke out he sent a message to the owners stating that he was saving the mill but was losing his wardrobe of which he was inordinately proud. By telegraph a speedy reply suggested that he save his wardrobe, but by all means to let the mill burn. About twelve miles east of Hamilton was Shekel's ranch, a stage station, where Mrs. Alice Shekel earned an enviable reputation for hospitality by insisting that no hungry man be turned away from her door. James R. Withington owned the Withington Hotel in Hamilton, at the time one of the most expensive structures in the state, and also had large ranching interest in the White River Valley. He went into the wholesale meat business in San Francisco, but ran into stiff competition from the well-known cattle and land barons, Miller and Lux, who had built an empire in California and Nevada. Everyone knew Uncle Tom Starr who carried the mail on snowshoes between Hamilton, Treasure City, and Eberhardt during the winter season. Adam Johnson was also well known for introducing the first ten-cent piece into camp, chiefly because the use of this small coin was unpopular — nothing smaller than a silver quarter had previously been used.[180]

When the British ceased operations on Treasure Hill in 1885, Drake, who had an option on the Monitor Mill and Mining Company at Taylor,

Nevada, about sixty miles from Hamilton on the western slope of the Schellburn Range, urged the company to take over this property rather than go into liquidation. The plant at the mine included a sorting room, blacksmith shop, and assay office. There was also a boarding house for the miners. The mill was located on Steptoe Creek, eight miles from the mine, where water was sufficient for operations.[181] Resolutions were passed at the annual meeting of the Eberhardt Company to buy this property for £36,000. It was also agreed that an effort would be made to raise £50,000 by debentures. The Monitor Mine was working, according to reports, at a profit of £18,000 a year.

A new company, known as the Eberhardt and Monitor Company, Limited, was registered in November, 1885, with a capital of £260,000. Within six months almost two hundred thousand £1 shares had been subscribed and £167,375 paid up.[183] Shares in the company fluctuated wildly during 1886, being worth only thirty-two cents at the beginning of the year, rising as high as $3.75 before dropping back to $2.00 by the year's end.[184]

In 1886, the shareholders learned they had purchased a worked-out mine. Ore had been found at the Monitor in pockets and chambers, like that at Treasure Hill, and in the thirteen months while the company was being legally reconstructed and the purchase of the Monitor Mine completed, the vendors had taken out all the worthwhile ore. The company was once again an exploring outfit. What was worse, Drake and Applegarth had left the company without warning and gone to operate the Palmarejo Mining Company, Limited, at Chinipas, Mexico. Frank Drake had again placed company affairs in the hands of his brother, Oliver. Reports from Nevada indicated that the Monitor Mill was barely paying its way. The chairman defended Drake's action by saying, "None of us who recollect what he said in this room or at the offices of the company can believe for one moment, in view of the large interest he holds at the present time, that he otherwise than expected very good results would accrue from the purchase of the Monitor property. . . . He had no idea the owners were taking such undue advantage of the delay and taking for the time being all the good ore they could reach." Embittered shareholders, particularly the newer investors, insisted that Drake's "retirement" looked to them like running away, that he had abandoned the management after recommending to the company a property whose value had been misrepresented.[185]

No matter how dark the future looked, the two British enterprises in White Pine County were reluctant to give up their corporate existence. For a half-dozen years the Consolidated Mining Company continued to invest in mining ventures all over the world. There was never an annual meeting that the taxes on the old South Aurora property and the Stanford Mill did not come up for discussion. A winding-up process was begun in 1886 and a "New" Consolidated Mining Company continued the search. Liquidation was never legally completed, but in 1905 the company's name was stricken from the official Register.[186]

In 1888, the Eberhardt and Monitor Company announced that the mill at Taylor had been shut down. Oliver Drake had also left the employ of the company and the new superintendent was William Miles Read. Read, born in England in 1849, had come to the United States at the close of the Civil War and made his way to Hamilton in 1873. Small in stature — five feet six inches — he had black, curly hair and deep-set dark-brown eyes and possessed the arrogance of the "little man" and the pride and dignity of an Englishman. Frank Drake, huge and robust, as burly as the typical miner of Nevada, was attracted to the newcomer and introduced him to a niece, Rose, whom he subsequently married.[187] Drake employed Read as a miner for six years and then made him general mine foreman from 1879 to 1887. The former superintendent recommended his protegé highly: "I always found him to be a man of good ability faithful to business, truthfull & honest in all his dealings and strictly Temperate in his habits."[188] The London directors were pleased to place an Englishman in charge of their property, particularly since he owned five thousand shares in the company. The Drake brothers held 7,500 shares between them and Edward Applegarth had over 10,000 shares so their interest in the enterprise was also guaranteed.[189] Moreover, the Drakes had signed over to the company two claims on Treasure Hill, the John Wild North and the John Wild South.[190]

In April, 1888, the New Eberhardt Company, Limited, was organized with a capital of £75,000, divided into 250,000 ordinary shares and 50,000 preference shares of five shillings each. Originally the company had expected to give old shareholders preferential holdings, but the public would not subscribe a shilling for common shares. In the end, only debenture holders got preference shares along with a few new subscribers, and shareholders in the old company took the ordinary shares. Four months later no one but seven signers of the Memorandum of Association attended the first statutory meeting.[191] Enough new capital was finally raised to resume work on the Monitor Mine and the directors hoped to start up the mill again. When the shareholders were called together in December by the new chairman, Francis Joseph Blandon, *The Financial World* of London reported the proceedings in a satirical fashion in an article entitled, "Mr. Blandon to the Rescue!"

> Mr. Blandon was in excellent form on Wednesday. He had much
> to say about the reconstruction of the Eberhardt and Monitor and
> he said it in that convincing manner of his which carried all
> before him. Perhaps, if we were disposed to be critical, we might
> say, he, to some extent, over-elaborated his statements, but share-
> holders are not disposed to find fault with minuteness of explana-
> tion. . . . More can scarcely be said.[192]

The journal questioned the logic of going to the expense of registering
a new company. The next year Blandon was dead.

At Taylor, Read worked hard to make the Monitor Mine pay.
Development continued and a limited amount of paying ore was made
available. However, operational expense for 1889 was £2,445 greater
than the income from silver. Employing rigid economy, Read continued
his efforts in 1890. In April the output amounted to $9,000 and the
expenses were only $4,500 but there were few months of the year so
profitable, and when the balance sheet was prepared at the end of June
the company was again in the red by £791. Read was paid £1,000
a year as general manager.[193] For her part, Rose Read purchased tea,
flour, sugar, salt, rice, and barley in hundred-pound sacks from the
Mormon wagon trains that made infrequent trips from the settlement at
Saint George, and supervised the meals for the men's dormitories. She
made her own soap, cast her own candles, and taught her two daughters
their lessons.[194]

In 1891, Read recommended that the company reduce the force to
a half-dozen men at the Monitor so the old Eberhardt and Aurora workings on Treasure Hill could be reopened. By working the properties
simultaneously the volume of silver was increased but not as rapidly as
the price for the metal declined. The company secretary, Alfred Oxenford, notified Read that the London directors relied completely on
his judgment.

> We have so much confidence in anything you deem necessary to
> do that we really thought it scarcely worth our while to make any
> observations. . . . We have been & are very pleased at your straight-
> forward manly reports which you send us weekly & it is very
> pleasant for us to feel that we can so completely rely upon them.
> You never go into high flown language & if you do err it is always
> on the side of prudence and consistency. — What more can I say
> . . . If anyone should know the *Eberhardt* you should. I have spoken!

When the superintendent periodically explained his urgent need for
money, the company informed him, "Of course you want men & equally
of course you must have them, but as regards the funds, we must cut our
coat according to our cloth." An agreement was reached whereby
expenditures would be limited to $4,000 a month and all went well as

long as Read could process enough ore to pay expenses. The company expressed admiration for his "steady and dogged perseverance." In 1892, Read proposed that he return to England to place the company's situation before the directors, and the company secretary assured him, "The attitude you have assumed toward the Board and through them to the Shareholders has (I speak honestly) materially altered the aspect of affairs & if, or I shall say when, you visit England & meet them your reception will be most flattering, of this I am fully convinced." In London, Read recommended that the company call up another shilling on each share for development work, and if nothing were found to shut down the Eberhardt. When he returned to Nevada, the company secretary wrote him:

> We have been up to the present quite unable to meet your requirements much to our annoyance. There are a great many laggards who have not paid their call . . . if shareholders will not pay the position remains practically unchanged. . . . But the position is an extremely uncomfortable one & the lack of good or encouraging news makes matters worse. . . . I am glad, though, that everyone is unanimous in their praise of your management & the failure is simply attributed to undertaking the impossible & not from any lack of knowledge or skill on your part . . . but these constant calls make the shareholders very irritable.[195]

In June, 1893, all work in Nevada was suspended. The New Eberhardt Company voted to leave the United States and try its luck with a gold mine near Barberton in the Transvaal. The directors paid for this Thistle Reef Gold Mining property in shares, nominally worth £75,000, with the understanding that the uncalled capital of the company, about £3,000, would be sufficient to put the mine on a paying basis.[196] The title to all the company properties in Nevada were assigned to William Read by a new Board of Directors with the understanding that, as company agent, he would dispose of them. No buyers could be found, and by 1896 the Thistle Consolidated Mines, Limited, had also failed. The company by this time owed Read $9,600 salary and the shareholders passed a resolution transferring to him "certain tracts, pieces or parcels of mining lands, mining properties, machinery, buildings and all other property of the corporation in the County of White Pine, Nevada."[197] In addition, Read also received the Thistle Reef property in South Africa. The secretary wrote him, "I am, in conclusion, desired by my Board to again convey to you our hearty thanks for your forebearance & we trust we may meet again in some future business that may be in every way as successful as the Eberhardt matters have been unsuccessful."[198] Read devoted much of the remainder of his life trying to interest capitalists in his White Pine holdings, hoping for a mining revival.[199] Early in the twentieth century, another English syndicate approached him about inspecting a property and advising them relative to its purchase. To this he agreed and stated:

> I am very glad to learn that English investors are again turning to America. Without doubt there are as good opportunities for investment in good mining properties in this country as in any country under the sun.[200]

THE MINING CAMP
AND THE FRONTIER

The pragmatic optimism of William Read, in spite of the years of hardship and adversity he and his British colleagues had experienced in eastern Nevada, was in the true spirit of the American Dream. In its simplest form, this dream was based on the right of every individual to pursue happiness and to seek success as he envisioned, it coupled with an optimistic belief that the quest would end in triumph. No group shared this vision, or believed stronger in the idea of progress, than those who participated in the frontier experience in the United States. Many scholars have noted that the myth of western opportunity was far greater than the reality of experience. Nevertheless, the belief in the glorious opportunities offered by the frontier was as strong a motivating force in luring men ever westward as the most satisfying experience might have been. The miners in the foothills of the Sierra, in the Rocky Mountains, or in the Great Basin believed in the Myth of the Pot of Gold at the end of the trail in the same way that the pioneer farmer moving into the prairies and the plains sought the Garden of Eden. Just as the hill or valley in the next county always looked a little greener or a little more fertile to the western-moving agrarian family so the ledge or outcropping in the mountain or plateau just at the edge of the horizon seemed certain to contain precious metal to the prospector. Time and time again the romantic dream of the frontier was destroyed by the harsh reality experienced by the fur trader, the cattleman, and the lumberjack as well as by the miner and the farmer, but the dream persisted. Undoubtedly it was a vital motivating force in the individual's determination to begin again, "the perennial rebirth" of experience that

[213]

Frederick Jackson Turner considered so essential to an understanding of the frontier. Certainly William Read and his kind expressed the spirit of frontiersmen through the ages.

————— •━• • —————

Many Americans residing in Western communities established as a result of the discovery of mineral wealth had difficulty in sustaining their optimistic dreams in the years immediately preceding the discovery at Treasure Hill. California's flush period following the first great gold rush of 1849 was over. The hordes of men, inexperienced in mining, who had arrived by sea or by overland routes in the first exciting season following the discovery, quickly filed claims to all the available placer deposits. Only a few were successful in locating a profitable claim. Within two seasons the amount of gold taken from these placers began to decline. Quartz deposits were being uncovered from time to time but their development took capital for machinery and also technical skill. Men who had operated alone or in partnership were soon forced to work as hired laborers for a daily wage under the supervision of a company foreman who, in turn, took orders from capitalists in San Francisco. Wages were higher than in the East but so were prices. As mining declined in the 1850's, the state had a tremendous surplus population and job competition forced wages steadily downward each year. Many men abandoned mining and flocked to the cities and towns to find employment. Industry and business were not sufficiently developed to absorb them and grave problems associated with unemployment, poverty, and unrest developed. Enterprising men eager to initiate business found that capital was scarce and interest rates almost prohibitive. Ranching and agriculture likewise provided no relief for the unemployed miner. Cattle operations required large acreages of land that the displaced miner had no hope of acquiring. Those from the East attempting to farm as their past experiences dictated learned that California soil conditions were quite different. The vagaries of nature, with alternating seasons of drought and flood between 1855–64, led to distress and discouragement. Agriculture was in the experimental stage and only by trial and error did the farmer learn what products could be successfully grown. Irrigation methods were inadequate and water rights not clearly defined by the law or the courts. As a capstone to his problem, the rancher or agriculturalist knew that his title to land, more often than not, was uncertain. So discontent prevailed in rural California as well as in the towns. Thus, as the economic position of the miners worsened, many found that there was no place for them in California.

Pacific Coast residents blamed much of their economic plight in the 1850's on their geographic isolation. Like frontiersmen everywhere in the West they appealed to the United States government to alleviate their distress by subsidizing improved communication and transportation. When Congress substituted a wagon road construction program across the Trans-Mississippi West in lieu of the requested railroad, and on the eve of the Civil War reduced overland mail facilities because of heavy financial losses, politicians began to talk of splitting the union three ways, rather than two, and creating a Pacific Republic. To insure loyalty and to reward the founders of the Republican Party in California, a grateful administration approved of the Pacific Railroad project. At last, the capital desperately needed to develop the state was forthcoming. Even before the railroad project was completed, however, residents realized that this was a mixed blessing. For with capital came monopoly and potential exploitation. Regimented economical labor in the form of Chinese coolies depressed the labor market. Many thoughtful Californians had come to the conclusion by the close of the Civil War that the real need of the state was capital and subsidy for mining development comparable to that dedicated to transportation.

The second major mining rush in the Trans-Mississippi West, that of 1859 to the Colorado Rockies, in its initial stages followed the pattern established in California. The placer deposits were exhausted after the first three or four summers of work. As these declined, lode mines were located and miners soon realized that their great opportunity was in this type of operation because the narrow, rugged canyons of Colorado were limited and difficult of access for gravel washing. The ores on the ledges as they went deeper were found to be in chemical combination with sulphides. Old processes of pulverizing the ore in a stamp mill and of using mercury to reclaim the precious metal by amalgamation did not work. No practical process was known whereby the gold could be reclaimed from these refractory ores. The Colorado mining industry appeared to be doomed in the mid-sixties unless science and technology came to the rescue. Experimentation was costly, so the cry, as in California, was for investment capital. The period 1865–68, represented the nadir of mining in the central Rockies. The economy was further depressed by agricultural distress resulting from grasshopper plagues and crop failure. The Colorado farmer was no better off than the Californian. A widespread Indian War drained manpower away from productive enterprise in these same years, threatened supply lines, and caused prices for essential commodities to soar. A shattering blow was the decision of the transcontinental railroad to by-pass Colorado and follow a route to the north across southern Wyoming. Many dreams of happiness, success, and progress were dispelled like bubbles bursting in the air.

Numerous Californians and Coloradans had wandered into the Pacific Northwest between 1860–64. Gold was found on a tributary of the Snake River in 1860; the next season there was a rush of men to the site, more than could be accommodated, so the surplus moved further south and made another discovery late in that same season. Next year brought an overwhelming rush, additional disappointment, a move southward to another tributary and a new discovery. A systematic three-year cycle of discovery, rush, and decline seemed to characterize each camp. They came so rapidly one after the other that the cycles overlapped, not having time to run their three-year course before another had begun. The excitement was largely over within four years. Across the mountains in future Montana disappointed prospectors from the Snake River discovered gold in 1862. Each successive season was characterized by a new discovery and the growth of a boom town that became the temporary capital. Here too the productive gold regions were quickly located and although prospectors moved out in all directions from Helena, after its discovery in 1864, no new field was located. Miners in the Northwest pondered their fate just like those in California and Colorado.

Simultaneous in time with the discovery of gold in Colorado, rich ore was found on the eastern slope of the Sierra at the Washoe mines of Nevada. The Comstock Lode was extensive and rich but costly to mine. In short order, San Francisco capitalists provided the money necessary to reclaim the metal from the potential mines, much to the disgust of Californians who felt that the funds for mining development should be used within the state. A four-year mining boom ensued, but when no new discoveries were made, this district, like all the rest, began to decline. Share values in the Comstock mining companies dropped precipitously, and when shareholders refused to pay calls to finance further exploratory work, many of these corporations collapsed. California bankers for a time tried to check the declining conditions by generous loans but when the depression hit in earnest in the two years following the close of the Civil War foreclosure became necessary. Meanwhile, miners who had been squeezed out of the Comstock by lack of capital to invest and later by declining wages had spread out over the state forming new mining districts. In these newer areas there was a clamor for investment capital to erect stamp mills, improve mining machinery, and in some cases, build smelters. On the Comstock, money was desperately needed to dig deeper, to shore up the underground caverns with timbers where the gold and silver had been extracted, and to drain the mines.

From the discovery of gold in California, 1848, to the close of the Civil War, the mining industry of western America was in an immature extractive stage. In the mining seasons of 1866–68 the industry was at a standstill everywhere — in California, Idaho, Montana, and particularly in Colorado and Nevada — desperately in need of investment capital to pay workers' wages, to experiment with tunneling and draining mines, to build tramways and railroads connecting mines with mills, to erect experimental smelters, and to finance the work of mining engineers. With this capital the mining industry of the United States could advance to a new level of maturity. When capitalists of San Francisco, Chicago, and New York hesitated to take the risk, mining "captains" of the West turned their attention across the Atlantic Ocean and found an enthusiastic reception to their proposals for investment, particularly in London. At last, the western American miner and promoter had found others to share their dream of optimism and faith in the future.

————•————•—•———

Britain had secured her position as the leading creditor nation of the world by 1815, a place she was to occupy without question for a century. The migration of capital abroad became a distinctive feature of British finance and commerce in the nineteenth century. The limited demand for money at home was a determining factor in this development as well as the comparatively higher earnings on money in other areas such as the United States. The funds available represented not only the savings of the British people amassed over many generations, but the profits from trade and industry accumulated in the early decades of the nineteenth century. After 1815 the flow of British capital to America was continuous, but in some years the stream became a rushing torrent and in others it virtually dried up. Prior to the American Civil War, investment was an individual matter with investors seeking out and purchasing likely looking American securities, usually upon the advice of a banker or counselor. This capital went largely into banks, into bond issues floated by states for public works, and to a lesser extent into business enterprise. By mid-century English investors, overburdened with capital, were eagerly seeking suitable investment fields. The California gold discoveries, for example, had precipitated a rapid movement of capital to the Pacific Coast. The economic dislocation caused by the American Civil War forced a curtailment of British investment, but confidence in the economic future of the United States came back rapidly after 1865. A major factor was the weakened position of the dollar and the exceptionally favorable exchange rate for British sterling. Channeling of funds overseas was no longer an individual matter, for the investor now used the joint-stock company and investment trust with headquarters in England or in Scotland.

The years 1870–73 were characterized by prosperity, and the British rapidly expanded their investment in the capital goods industries and the railroads of both continental Europe and the United States. Railroad builders, eager to tap the resources of vast underdeveloped regions with cheap means of transport, set off a great construction boom that became the backbone of the world economy. By supplying the railroad iron, rolling stock, and construction capital the British largely underwrote this vast expansion. The nation also increased the volume of loans to governments in Europe, South America, and the United States. The British export of capital, on a world-wide basis, had been increasing slowly each year since 1865 and stood at £28 million in 1870. The sudden boom of the next three years pushed the figure to an all-time high in 1872, £76 million, representing a two and one-half increase within this short span of time. There was a slight tapering off in 1873 to £63 million.

The great boom in the export of British capital was accompanied by the launching of many new overseas mining endeavors and by a frenzied speculation in the shares of this type of enterprise in London. The *nominal* capitalization of companies floated in London to explore prospects, purchase mining claims, and actually operate mines and mills reached an annual volume of approximately £40 million in 1872 and 1873. These years were hereafter referred to as the period of the "first mining mania." Observing the rising tide of enthusiasm for mining securities, the London *Times* began a campaign to show the dangers involved in haphazard investment. Its effort apparently checked few investors and only seemed to stimulate protests from mining journals throughout the world.

The sudden boom in British mining investments was felt in western America as elsewhere. At the beginning of 1870, San Francisco's *Scientific Press* noted, "Our mines are attracting the attention of European capitalists, and already several have passed into the hands of English companies. This is but a beginning; we have plenty of developed workings, which, if properly managed, would surely yield profits large enough to attract foreign capital." While the London *Mining Journal* and the *Mining World* debated the question of whether or not American mining would pay, investors did not wait. In a four-year period, 1870–73, sixty-seven mining companies, capitalized at about £14 million were registered to operate in the western United States outside of California. In addition, twenty-seven companies, representing over £4 million were active in that state, making a grand total of ninety-four companies capitalized at £18 million. A Pacific coast magazine explained the opportunity: "There are many good mines west of the Rocky Mountains, which, by enlarging their works and developing further, might be made to pay handsomely, but which, from the scarcity of capital and the high rates of interest here, are unable to make the requisite improvements, but which in European cities, where there is often a plethora of money, and a less rate of interest prevails, would be the best investment that could be presented to the notice of capitalists." Although California proved to be the favorite field in western America for British mining investments in these years, Nevada was not far behind in company registrations. Colorado and Utah, in third and fourth place, could boast of twenty or more British enterprises working within their borders. There was also some British mining capital in Arizona, Idaho, and Wyoming during the period. Several companies launched in western America between 1870–73 were among the largest and most notorious sponsored by the British throughout the world in the nineteenth century. In California, the Sierra Buttes Mining Company, Limited, and the Plumas Eureka Mining Company, Limited, operated the "show mines" of the state, noted for efficient management, the production of profits, and the consistent declaration of dividends. The Richmond Consolidated Mining Company, Limited, in Nevada paid annual dividends from 1872 through 1895 representing a 324 per cent return on the capital invested. On the other hand, the Emma Silver Mining Company, Limited, of Utah ended in international scandal, and there were notorious failures and reorganizations of the Flagstaff Silver Mining Company and the Last Chance Silver Mining Company of that same state and the Ruby Consolidated Mining Company of Nevada. Whatever was done seemed to be done on a grand scale.

During the Civil War, British capitalists were first attracted to mining investments in the American West through promoters of Nevada mines. On the basis of the success of the Washoe mines in western Nevada three joint-stock enterprises were registered in 1864–65 to operate properties in the Reese River District. Each of these companies was incorporated at £100,000, but only the first one launched actually engaged in mining. In 1867, a Lander City Silver Mining Company, Limited, was also legally established, but did no work. Significant British mining operations in Nevada did not start until 1869 when the Battle Mountain Mining Company, Limited, and the Pacific Mining Company, Limited, raised funds in London. Then the news of the new

district at White Pine became known in British financial circles and
four Nevada mining companies, including the Eberhardt and Aurora
Mining Company, Limited, and the South Aurora Silver Mining Coming Company, Limited, came into being in 1870. The number of British
companies beginning work in Nevada during 1871 doubled to eight.
Thus, British investment in Nevada mining followed the pattern established for all the western United States, or for that matter, the entire
world. The events at Treasure Hill, in the White Pine Mining District of
eastern Nevada, were part and parcel of British overseas investment
in the 1870's.

Before the British took over in White Pine the district had already
run through most of the stages in the life cycle of a typical western
mining community. First of all there had been a chance discovery in
a little known and isolated area in 1867. As is so often the case, the
region was difficult in topography and climate. The original discoverers
attempted to conceal their find and to organize mining district laws to
favor the first arrivals. Then the news leaked out concerning the discovery and soon exaggerated rumors of great wealth were scattered far
and wide. This led to the usual rush during the summer months of the
year following the discovery. Inevitably more men came than could be
accommodated and there was disappointment and hardship. Everything
was overdone including the registration of claims, promotion of real
estate, development of stage lines, stocking stores with equipment and
food supplies, building saloons, even patronizing the theater and prize
fights. As soon as the crowd arrived, the discoverers sold their claims
to capitalists for a pittance of their value. Others who had worthless
properties took advantage of the inexperienced to unload. Disappointed
miners usually fanned out in all directions from the center of the discovery, and the surrounding country was thoroughly prospected. By
the third season the value of the district was rather well established.
If the boom had been based on exaggerated reports, collapse came
quickly to the area where the cost of living was high. Merchants who
had overstocked were ruined.

After the flush times of prosperity, the mining district faced one
of two fates. It might survive a period of decline, even depression, and
then a new discovery accompanied with greater capital investment and
improved technology might lead to a revival and stability for a longer
span of years. This is what happened at the Comstock Lode in western
Nevada. On the other hand, a district was more often faced with slow
attrition leading to final collapse, a time characterized by continuous
shrinkage of population while small-time operators struggled to make a

living as the output of the mines declined. At Treasure Hill the choice of these two fates was not as pronounced as usual because of the role of the British. They provided capital and technical skill to revive the mining industry of eastern Nevada after the cycle had passed the boom stage and was well on the way toward decline. For a while it appeared that the district would have a future of prosperity and stability experienced by the vast minority of western mining areas. But the new discovery was not forthcoming so the British could not control the cycle that went on its inexorable course toward collapse and the ultimate fate of most mining communities, that of becoming a ghost town and a future tourist attraction.

When the contribution of the British to the mining industry of western America is assessed, examples can be found to illustrate every characteristic of business enterprise in the nineteenth century. On the positive side there can be seen wise investment and use of capital, pioneering endeavor in the fields of science and technology, encouragement to mining engineering, capable management both in London and at the mine site, enlightened labor policies, profits justly earned, and dividends systematically distributed to shareholders. On the negative side there was ample evidence of the selling of worthless properties, collusion between vendors and promoters to bilk the investing public, overcapitalization, extravagance, mismanagement, and even proven fraud. One fact stands out above all others. The British invested far more in the Mining Kingdom of western America than they ever reaped in profits. Much of the loss came from enterprises that never had a chance of success. Many legitimate companies, like the major operations in White Pine, that worked systematically and doggedly for success failed to achieve it. Even those British mining companies that located paying gravel or veins, few in number, spent the largest portion of their profits paying the wages of miners and millmen, buying supplies and equipment and experimenting with latest technological improvements rather than returning investment capital to their British stockholders. Not more than ten or twelve companies out of the five or six hundred registered to operate mining enterprises in the American West between 1865–1900 returned the capital invested by British stockholders in the form of dividends. Many of those that succeeded in doing so took over twenty years in the process.

As in the case of Treasure Hill, 95 per cent of British mining efforts throughout the world brought no financial return. The historian finds difficulty in explaining why well-informed people with an urban, commercial, and industrial background were repeatedly drawn into impossible ventures. The excitement of a plunge into gold and silver has always appealed to men, and will continue to do so, no matter how risky the project, how dim and sketchy the plans for development, or how grandiose the prospectus. The discovery of one good mine apparently counterbalanced the hundreds of failures in the eyes of the venturesome. Perhaps no better explanation of the events in White Pine County, Nevada, can be made than the persistence of the British coupled with the persistence of the American Dream both at home and abroad. This dream of opportunity and optimism was not the exclusive property of the western pioneer. It had become a part of the national mind and spirit.

———•➤•———

As one reflects on the events at Treasure Hill, the foremost question that comes to mind is the place of the Western mining district in the frontier experience of the nation. To what extent were the activities and attitudes of the men and women who came to eastern Nevada as the result of the discovery of silver like those in mining districts elsewhere? Were they similar or dissimilar to other types that were a part of the westward movement? Above all else, the frontiersman's life and institutions were shaped by the impact of natural conditions, the environment in which he found himself. Certainly this was true of the miner. Just as the farmer sought soil that would produce better crops, the miner sought mountain streams, quartz ledges, and underground gravel beds that would produce precious metal. The geographic environment dominated the movements of the western miner and provided the only excuse for a mining rush or the organization of a mining community. Certainly in the initial stages of any mining district, and for the entire life of some, the central theme was that of man against nature rather than man against man. As we have noted, most mining districts were isolated and in unknown country. The terrain was usually difficult. In the vast majority of mining districts operations were of a seasonal nature because of the bitter cold of winter. The White Pine Mining District possessed all the environmental chracteristics of the typical mining area but to an exaggerated degree. Few discoveries of gold and silver were made at such a high elevation, seldom did the temperatures of winter drop so low, or the winds blow so long and hard, or the working season last so briefly. Where on the frontier, one may ask, did the natural geographic conditions and the environment play a more dominant role than in the mining camp?

The frontier was an area where men from many nations and many sections of the United States formed a new society enriched by their experiences and institutions elsewhere. So it was with the western mining camp. By the time of the Treasure Hill discovery, the experiences in California, Colorado, Idaho, Montana, and western Nevada were all in the past, and even the most inexperienced were familiar with basic mining methods as well as procedures for registering claims and establishing a mining district. However, innovation based upon expediency was more highly valued on the frontier than tradition. Miners, like those at White Pine, did not hesitate to modify claims law or mining district regulations to suit the changing situation and times. Their needs also stimulated invention. For example, as long as rich silver chloride ore was being mined, the stamp mills provided the only need for processing, but when it was learned that low-grade refractory ores containing some sulphides might provide economic stability for the community, experimentation with reduction furnaces was immediately begun in Hamilton. When winter snows made the transport of ore from mine to mill difficult, if not impossible, the most elaborate tramway in western America was constructed to overcome the problem. When the conveyor wires sagged dumping the ore, a phenomenon beyond the previous experience of the European inventor and engineer, the cause was found to be rapid changes in temperature at the high elevation affecting the expansion and contraction of the metal wires, and steps were taken to remedy the situation. If White Pine District was typical, the miner like other frontiersmen profited by experience of the past in distant places, but also encouraged experimentation, innovation, and invention.

Scholars are sharply divided over the extent to which the frontier produced individualism. The mining camp provides an excellent example of the cause of their division. The vast majority of men arriving in any mining district were egoistic, hoping to operate alone, although some arrived with one or more partners with whom they had worked in the past. Individualism appears to have broken down rather rapidly with two to four men binding themselves together to prospect, register, and work claims, build a shelter and provide subsistence. In many cases a careful division of labor was agreed upon. In an all-male society, these cooperative arrangements may have served as a substitute for the family unit on the frontier farm. As society became more complex this spirit of cooperation was extended to include the law enforcement group, the fire-fighting company, and even the miners' union. Moreover, the frontiersman never let his individualism stand in the way of seeking government assistance for projects too large either for individual or cooperative efforts. Because of their isolation, mining camps seemed more concerned over improvement of transportation and communication than any other aspect of government aid, and like the miners in eastern Nevada they wanted improved mail facilities and government subsidies for telegraph and railroad construction. A careful examination of the mining camp causes one to question the ability of the average miner to express his individualism very long after arriving in a new district. In the social and recreational field he appears to have been a free agent, in politics to a lesser extent, but in his paramount economic concerns, militant individualism was likely to lead to disaster.

The frontier has been glorified as a stimulus to democracy both in the West and throughout the nation. Some writers have pictured the mining camp as a vital expression of this democracy with its institutions of the camp meeting, the equality pervading the mining district law and vigilante action. Emphasis is placed upon the leveling influence of migration. Yet from the very beginning class lines and conflicts based upon social status and economic position not only prevailed but also were commented upon. There were the workers as opposed to the idlers on the streets, the ridiculed prostitutes vs. the glorified dramatic artists, and later the labor unions vs. the capitalists. When district regulations were replaced by county and municipal government, even the democracy of the political process broke down. Another national characteristic, the apathy of the average citizen, dominated the political scene. Refusing to pay taxes except under duress, neglecting to vote unless vital patronage matters were at stake, the average miner appears to have been politically passive. He was dedicated to amassing material wealth and largely unconcerned about democratic theory, ideals, and institutions. His democracy did not concern itself with the rights of minority groups. The historian needs to take a second look at the democracy of the mining camp, examining not only the process and institutions associated with the rush and early seasons, but also developments on the democratic front when the district began to decline and gain stability. Fairness demands that the democracy of the mining camp be compared to its existence elsewhere in the late nineteenth century rather than to democracy of the present.

The miner, like frontiersmen and the rest of the United States citizens, was materialistic and had a tendency to judge success by the attainment of wealth. He had a great respect for personal property as evidenced by claim registrations, and he believed in the sanctity of contracts. At the same time he was guilty of exploitive wastefulness both of natural and human resources. The richest gravel or the most valuable pockets of ore were quickly taken without regard for the systematic working of a claim. If he was secure in his property through local district regulations, the average miner was too impatient to take advantage of the federal land laws that guaranteed the title to his claim beginning in 1866. Often times wastefulness was a corollary to the materialism of the frontier as expressed by the fur trader, the buffalo hunter, the lumberman, and the pioneer farmer as well as by the miner.

Early enthusiasts for the frontier emphasized its leveling influences, the processes whereby the customs and institutions of a foreign land were shed and from which a composite Americanism emerged. The frontier was often described as a melting pot where the diverse became uniform. Out of this amalgamation there supposedly emerged patriotism and a nationalistic outlook. This was not the experience in the typical mining camp. In fact, the diverse role played by various national groups contributed to the social charm of the community, aided in explaining its political alignments, and influenced its pattern of economic growth. Although national pride and patriotism were demonstrated in the celebration of American Independence Day and Washington's Birthday, the Irish never failed to recognize the natal day of St. Patrick with greater enthusiasm. While special American holidays like Thanksgiving were observed, Christmas provided an opportunity to preserve the traditions of the old country, whether it was Scandinavia, Germany, or another. When Franco-German relations became tense, the conduct of the German miners and French shopkeepers in the western mining camps was such that in the twentieth century they would have been labelled hyphenated-Americans. In most of the miners' unions the organizing ability and leadership of the Irish was as apparent as it was elsewhere in the labor movement of the nineteenth century. Certainly the process of assimilation failed to work to the slightest degree in the western mining camps where the Chinese and Mexicans remained forever apart from the rest of society. Nor does one sense that the miners thought of themselves as an expression of national expansion, of pushing forward the boundaries of the country with the same sense of mission that was felt by other frontier types, including the pioneer farmer.

The frontier experience has been credited with producing and nurturing personality traits that have come to be thought of as distinguishing the American. The nation and its people are a restless lot. Certainly the frontiersman, and particularly the miner, were highly mobile. Most of the men arriving in a mining camp, like that at White Pine, had been on the move for years and the moment fortune frowned on them, they were not loath to move again. The nation and the frontier are considered youthful and hardy. The available evidence indicates, however, that the age span in the later mining rushes, after those of 1849 and 1859, was much greater than has been assumed. This may also have been true of other frontier groups like the cattlemen, sheepherders, and pioneer farmers. Although miners displayed physical vigor and strength and as a group enjoyed good health, the limited evidence suggests that many of them suffered from exhaustion and exposure and endured chronic illness. The frontier demanded physical stamina but it was not always characterized by robust health. The frontiersman was also described as coarse, lawless and rambunctious. The miner was probably no exception for he went along with a lax enforcement of the law except in times of crisis when the pendulum was apt to swing rapidly in the direction of vigilante action. White Pine miners engaged in informal law enforcement, but never went so far as to establish a vigilance committee to take the law into their own hands. The crudeness of the miners' manners expressed itself most clearly in hours of recreation at contests of physical skill and endurance and at the theater. The frontiersman is often characterized as uncultured and anti-intellectual. In his endeavor to obtain economic security, he neglected aesthetic and intellectual pursuits. Once material success was achieved, he used his means to support literary societies, libraries, schools, and camp meetings that were a part of most western communities, although he may not have had a genuine appreciation of much that he financially supported as a matter of civic pride. The typical mining camp also backed the newspaper, the magazine vendor, and the dramatic artists far more strongly than the financial position of its inhabitants justified. In most mining districts there was greater support for the church and school than in White Pine. Certainly the few married women in the mining camp struggled for refinement and culture in the home and community to the same extent, and in doing so overcame hardships as great as those facing the pioneer farm wife. Both the miner and other frontiermen undoubtedly lacked sophistication, and past experiences may have left them uncultured, but one questions the judgment that they were basically anti-intellectual any more than the average American.

Miners not only shared the character traits of other frontiersmen, but had other attributes that set them apart. Although lacking a social conscience that directed consistent humanitarian endeavor, a community crisis or personal tragedy could produce unbelievable generosity. While taxes to support public institutions might be ignored, funds for a hospital or jail might be raised in a few hours by a public subscription. Widows and orphans were not neglected. Nor was a fellow miner, suffering from ill-fate or accident, forgotten. The mining camp was a center of emotional tension quite unlike anything experienced on the agrarian frontier and could burst forth in a humanitarian crusade or in labor violence. Although frontiersmen everywhere had a tendency to come to each other's aid in adversity, there was an exceptional form of fraternalism present in the mining camp. The brotherhood of man found organized expression in the fire-fighting companies, the labor union, and most of all in the Odd Fellows and Masonic lodges. This comradeship in life followed in death for the contemporary observer of any ghost town cemetery near an abandoned mining town will note that lodge membership was second only to the family and national origin in determining the place where the miner's remains were deposited. Even if he were buried in the family plot or the section reserved for the Irish, Italians, or Slavs, the symbol of the miner's fraternal lodge was always recorded.

Recent students of the frontier have chided earlier scholars for failure to note the importance of the land speculator on the agrarian frontier. No doubt jobbers were always ahead of the farmers in the westward movement, buying up the best sites, so that the settler on arrival had either to pay the speculator's price or accept an inferior location. For every homesteader who secured his acreage from the government, six or seven purchased their farms from speculators. The mining camp had its counterpart. The speculators in mining claims, town lots, and in the shares of mining companies were also notorious and significant members of society here. Those who took a big risk, who gambled or speculated, were an ever-present, vital component of a frontier experience that was far more complicated than the "orderly process . . . marching single file westward," as it was once described.

At one time exponents of the frontier suggested that the presence of available land in the West served as a safety-valve of escape for those who were distressed because of economic adversity or social pressure. Although their ideas were not always explicitly expressed, these scholars left the impression that the poverty-stricken, urban worker of the East could find escape from industrial depression by migrating west and taking up a farm. The combined effect of this process was to lessen explosive tensions in the East. More recently historians have argued that the picture of the penniless worker's family migrating great distances westward in time of depression to take up farming for which they had no experience is a gross misconception. A move was far more costly than the family could afford. What migration that did occur was in periods of prosperity and not depression. It has also been suggested that labor troubles and social discontent in the East were not alleviated by western migration but that the pattern followed similar developments in Europe. As the West drew off population from the East, the numbers were more than replaced by additional immigrants from Europe. One scholar has argued that what discontent existed was agricultural rather than urban and has proved by census figures that there was more migration away from the farm to the small town than there was to the farm.

Although one may grant that the frontier did not provide a solution to the Eastern workers' problems, there is no doubt that the mining rush served as a safety valve for many men in the West. Time and time again when the miner's luck had run out and he was at the end of his resources, a new discovery, or even the rumor of a new find, gave him a new lease on life. The release here was undoubtedly more significant than that on the agrarian frontier because the miner, as a single man, needed far less for a stake than the frontier farming family. Nor did the miner's movements have anything to do with cycles of prosperity and depression except those of the small community of which he was a part. The miners of the West constantly believed that they would escape adversity, that a new El Dorado would be discovered producing wealth for all, and this persistent belief in a "safety-valve" was a strong psychological force motivating their actions. Many miners spent most of their adult life roaming from mining camp to mining camp. This process did not lead to individual or community stability. However, miners like all frontiersmen were opportunistic and optimistic above everything else, and many of them throughout most of their lives had a boundless belief in the future — in The American Dream.

CHAPTER NOTES

Chapter I: THE BEGINNING

[1] Myron Angel, *History of Nevada*, p. 660 ff.
[2] Transcript of Records of the Miners Meeting, Tuesday, October 10, 1865, By-Laws of the White Pine Mining District, *White Pine News*, February. 6, 1869.
[3] Hubert Howe Bancroft, *Works: History of Nevada, Colorado, and Wyoming*, XXV, pp. 227 ff; Dan M. McDonald, "White Pine County," in Sam P. Davis, *History of Nevada*, I, pp. 1042-1043; Francis Church Lincoln, *Mining Districts and Mineral Resources of Nevada*, p. 257; *Report of the Nevada State Mineralogist, 1867–1868*, p. 51; "White Pine," *Overland Monthly* (San Francisco), III, (September, 1869), pp. 209-210.
[4] Allen Cadwallader, *Map and Guide to the White Pine Mines and the Region of Country Adjacent in Eastern Nevada*, p. 3.
[5] W. S. Lorsh, "Mining at Hamilton, Nevada," *Mines and Minerals*, 29 (June, 1909), p. 521.
[6] "The Geographical Position of the White Pine District," *White Pine News*, December 26, 1868; Tagliabue and Barker, *Map of the White Pine Mines and the Regions Adjacent*, p. 3; Cadwallader, *op. cit.*, p. 5.
[7] Rodman Wilson Paul, *California Gold: The Beginning of Mining in the Far West*; Robert Glass Cleland, *A History of California: The American Period*, pp. 225-246.
[8] Paul, *op. cit.*, pp. 179-181; W. P. Morrell, *The Gold Rushes*, pp. 135-141; Charles W. Henderson, "Mining in Colorado," in James H. Baker and LeRoy R. Hafen, *History of Colorado*, II, pp. 525-572; Percy S. Fritz, *Colorado, the Centennial State*, pp. 103-120.
[9] William J. Trimble, *The Mining Advance into the Inland Empire*, pp. 62-84; Morrell, *op. cit.*, pp. 165-187.
[10] Joseph Wasson, *Bodie and Esmeralda*; Angel, *op. cit.*, pp. 414-416; Effie Mona Mack, *History of Nevada*, pp. 213-214.
[11] Angel, *op. cit.*, pp. 449-453; Bancroft, *op. cit.*, pp. 262-264.
[12] Angel, *op. cit.*, pp. 425-426, 461-465.
[13] Minutes of Meeting of the Miners in the White Pine District, Recorder's Office, Mohawk Canyon, July 20, 1867. *White Pine News*, February 6, 1869.
[14] Tagliabue and Barker, *op. cit.*, p. 19.
[15] *Report of the Nevada State Mineralogist, 1867–1868*, p. 52.
[16] San Francisco *Bulletin*, March 8, 1869. An account written by "Pilgrim" at Treasure City, Nevada, and reprinted in Cadwallader, *op. cit.*
[17] Bancroft, *op. cit.*, p. 278.

[18] No mining community has received as much attention from writers as the Washoe District of western Nevada. Mark Twain's *Roughing It* and Dan DeQuille's (William Wright) *The Big Bonanza* are both classics. Of particular value are the edition of Twain's book published by Rinehart and Company with an introduction by Rodman W. Paul and the Knopf edition of *The Big Bonanza* ably introduced by Oscar Lewis. Less known but equally valuable studies are Eliot Lord, *Comstock Mining and Miners* (Monographs of the United States Geological Survey, IV, Washington, 1883), and Charles H. Shinn, *The Story of the Mine*. Worthwhile information is also found in two books by George D. Lyman, *The Saga of the Comstock Lode* and *Ralston's Ring;* Oscar Lewis, *Silver Kings;* and Franklin Walker, *San Francisco's Literary Frontier*. Two volumes that must be used with care provide interesting ancedotal material: Miriam Michelson, *The Wonderlode of Silver and Gold* and Ethel Manter, *Rocket of the Comstock*.

[19] Albert S. Evans, *White Pine*, p. 45, quoting the *Alta California* (San Francisco), February 18, 1869.

Promoting the Boom

[20] Cadwallader, *op. cit.*, pp. 5-6; James D. Hague, "Mining Industry," in Clarence King, *United States Geological Exploration of the Fortieth Parallel*, III, p. 422.

[21] Bancroft, *op. cit.*, p. 278; Angel, *op. cit.*, p. 661.

[22] *Report of the Nevada State Mineralogist, 1867–1868*, pp. 52-53.

[23] *Alta California*, February 8, 1869.

[24] San Francisco *Bulletin*, March 8, 1869, quoted in Cadwallader, *op. cit.*

[25] Hague, "The Mining Indusrty," *loc. cit.*, p. 431; *Report of the Nevada State Mineralogist, 1867–1868*, p. 54; Evans, *op. cit.*, p. 10.

[26] Evans, *op. cit.*, p. 14.

[27] *White Pine News*, December 26, 1868.

[28] Hague, "Mining Industry," *loc. cit.*, pp. 426-427; Evans, *op. cit.*, p. 11; *Report of the Nevada State Mineralogist, 1867–1868*, p. 54.

[29] Evans, *op. cit.*, pp. 13-14.

[30] San Francisco *Bulletin*, March 8, 1869.

[31] *White Pine News*, January 23, 1869.

[32] *Report of the Nevada State Mineralogist, 1867–1868*, p. 55.

[33] Cadwallader, *op. cit.*, pp. 8-9; Evans, *op. cit.*, p. 10; *Report of the Nevada State Mineralogist, 1867–1868*, p. 56.

[34] *The Inland Empire* (Hamilton), May 29, 1869. Reprinted from the New York *Herald*.

[35] Hague, "Mining Industry," *loc. cit.*, pp. 424-425.

[36] *White Pine News*, December 26, 1868.

[37] Tagliabue and Barker, *op. cit.*, p. 20; Hague, "Mining Industry," *loc. cit.*, pp. 439-440.

[38] *Mining and Scientific Press*, XVIII (January 2, 1869), p. 5, quoting the *Alta California*.

[39] *Ibid.*, XVII, (December 5, 1868), p. 354.

[40] *Alta California*, February 8, 1869..

[41] Quoted in the *White Pine News*, April 8, 1869.

[42] *The Inland Empire*, April 7, 1869.

[43] Quoted in the *White Pine News*, February 13, 1869.

[44] March 6, 1869, quoted in *Mining and Scientific Press*, XVIII (March 20, 1869), p. 180.

[45] *Mining and Scientific Press*, XVIII (March 6, 1869), p. 157.

[46] *White Pine News*, March 20, 1869.

[47-48] *Mining and Scientific Press*, XVIII (February 27, 1869), p. 140; (March 20, 1869), p. 184; (April 3, 1869), p. 212.

[49] "Deserted Districts," *White Pine News*, March 27, 1869.

[50] "The Prospecting Mania," *ibid.*, March 13, 1869.

Routes to the Mines

[51] *White Pine News,* January 2, 1869.
[52] Cadwallader, *op. cit.,* pp. 21 ff.
[53] *Alta California,* February 8, 1869.
[54] Cadwallader, *op. cit.,* pp. 21 ff.
[55] *White Pine News,* March 13, 1869.
[56] Charles Gracey, "Early Days in Lincoln County," *First Biennial Report of the Nevada Historical Society, 1907-1908,* pp. 103-105.
[57] Quoted in the *Mining and Scientific Press,* XVIII (April 3, 1869), p. 212.
[58] Aaron D. Campton, "Experiences of a Nevada Pioneer," *Second Biennial Report of the Nevada Historical Society, 1909-1910,* p. 103.
[59-60] *Mining and Scientific Press,* XVIII (March 6, 1869), p. 157 and (March 27, 1869), p. 196.
[61] Quoted in *White Pine News,* March 27, 1869.

Boom Towns

[62] Evans, *op. cit.,* pp. 8-9.
[63] *White Pine News,* January 30, 1869; *The Inland Empire,* March 31 and August 22, 1869.
[64] Evans, *op. cit.,* p. 9; *The Inland Empire,* August 22, 1869.
[65] *The Inland Empire,* May 25, 1869; *White Pine News,* August 25, 1869.
[66] *The Inland Empire,* April 2, May 2, and May 20, 1869.
[67] *White Pine News,* June 3 and 14, 1869.
[68] *Alta California,* February 8, 1869.
[69] *The Inland Empire,* May 25, 1869.
[70] *White Pine News,* April 8, 1869; *The Inland Empire,* April 10, 1869.
[71] *White Pine News,* June 4, 1869.
[72] *Ibid.; White Pine Evening Telegraph,* June 4, 1869.
[73] *White Pine Evening Telegraph,* June 11, 1869.
[74] *A Compendium of the Ninth Census, June 1, 1870,* 42 Cong., 1 sess. (Washington, 1872), pp. 12, 18, 72 and 256.
[75] *Ibid.,* pp. 8 and 425. The contributions of various states to the mining population at White Pine are as follows: Nevada, 128; California, 364; New York, 701; Pennsylvania, 295; Ohio, 360; Illinois, 158. Many were also foreign born: Ireland, 971; England and Wales, 572; Scotland, 119; Canada, 454; Germany, 445; France, 91; Sweden and Norway, 82; Denmark, 43; Switzerland, 38; China, 292; Mexico, 51.

What of the Prospects?

[76] Evans, *op. cit.,* pp. 32 ff.
[77] *Report of the Nevada State Mineralogist, 1867-1868,* pp. 53 ff.
[78] XVII (October 31, 1868), p. 280. For a more technical and complete discussion, see "Ore Deposits and Metallurgy at White Pine," *ibid.,* (November 28, 1868), p. 345.
[79] Quoted in the *Mining and Scientific Press,* XVIII (February 13, 1869), p. 103.
[80] *Ibid.,* (April 20, 1869), p. 180.
[81] Virginia City *Enterprise,* March 18, 1869, quoted in *Mining and Scientific Press,* XVIII (March 27, 1869), p. 196.
[82] Nevada *Transcript,* March 18, 1869, quoted in *ibid.*
[83] Quoted in *White Pine News,* April 6, 1869.
[84] *Ibid.,* April 17, 1869.
[85] Idaho *Tidal Wave,* March 9, 1869, quoted in *Mining and Scientific Press,* XVIII, (March 27, 1869), p. 196.
[86] *The Inland Empire,* April 27, 1869.
[87] XVIII (April 17, 1869), p. 249 and (May 1, 1869), p. 286.
[88] April 17, 1869.
[89] Hague, "Mining Industry," *loc. cit.,* pp. 424-425.
[90] April 21, 1869.
[91] Quoted in the *White Pine News,* May 7, 1869.
[92] *Ibid.*

232   TREASURE HILL

[93-94] *The Inland Empire*, May 23 and June 26, 1869.
[95] XVIII (May 15, 1869), p. 312.
[96-97] *White Pine News*, May 1, 12, 13, June 8, August 19, and September 14. 1869.
[98] *White Pine Evening Telegram*, July 6, 1869.

Chapter II: THE FLOURISHING

[1] *Mining and Scientific Press*, XVIII (January 16, 1869), p. 33 and (March 20, 1869), p. 180; *White Pine News*, February 6, 13, 1869.
[2-4] *White Pine News*, February 13, March 13, April 3, 1869.
[5] March 17, 1869.
[6] *The Inland Empire*, March 31, 1869; *White Pine News*, April 5, 1869. The *Empire* stated that twenty-seven passengers left Virginia City, but the *News*, quoting the *Reveille*, reported twenty-one passengers when the group arrived in Austin.
[7] Quoted in *White Pine News*, April 8, 1869.
[8] *The Inland Empire*, April 16, 1869; *White Pine News*, April 13, 1869.
[9-10] *The Inland Empire*, March 27, April 10, 16, May 2, 1869.
[11] *White Pine News*, April 5, 1869.
[12-13] *The Inland Empire*, April 16, May 19, 1869.
[14] *White Pine News*, June 28, 1869.
[15] April 7, 1869.
[16-17] *White Pine News*, April 9, 14, 1869.
[18] *The Inland Empire*, April 18, 1869.
[19] April 16, 1869.
[20-22] *White Pine News*, April 12, 22, July 17, 1869.
[23-25] *The Inland Empire*, March 27, May 25, 1869.
[26] *White Pine News*, June 7, 1869.
[27] June 6, 1869.
[28-30] *White Pine News*, April 15, May 11, 13, 1869.
[31-32] May 12, 18, 1869.
[33] *White Pine Evening Telegram*, June 10, 18, 1869.
[34] *The Inland Empire*, September 1, 14, 1869.
[35-36] *White Pine News*, September 23, 28, 1869.
[37] *The Inland Empire*, October 20, 1869.
[38] *White Pine News*, October 9, 1869.
[39] *White Pine News*, September 2, 1869; *The Inland Empire*, December 12, 14, 1869.
[40] *The Inland Empire*, December 2, 1869; *White Pine News*, December 4, 1869.
[41-43] *White Pine News*, March 27, April 19, May 25, 1869.
[44] *The Inland Empire*, November 2, 1869.
[45-46] *White Pine News*, January 30, March 27, 1869.
[47-50] *The Inland Empire*, March 27, April 9, 10, 11, 16, 17, 18, 1869.
[51] Evans, *op. cit.*, pp. 18-19, 40.
[52] *White Pine News*, April 9, 1869.
[53] *The Inland Empire*, April 22, 1869.
[54-56] *White Pine News*, April 6, 9, May 7, 1869.
[57] *The Inland Empire*, June 2, 18, 1869.
[58] *White Pine Evening Telegram*, July 15, 1869.
[59-61] *The Inland Empire*, September 5, October 27, November 14, December 31, 1869.

Environment

[62-63] *White Pine News*, February 6, 13, 1869.
[64] *The Inland Empire*, March 27, 1869.
[65] *White Pine News*, February 27, 1869.
[66-70] *The Inland Empire*, April 4, 10, 18, May 4, 26, 1869.
[71] *White Pine News*, August 24, 1869.
[72] *The Inland Empire*, August 29, 30, 1869.
[73-74] *White Pine News*, September 21, 30, 1869.

[75-78] *The Inland Empire,* October 1, November 7, 16, 24, December 2, 21, 22, 29, 31, 1869.

[79-80] *White Pine News,* February 25, 1869, February 11, 1870.

[81-82] *The Inland Empire,* February 25, 1869, February 26, 1870.

[83] *White Pine News,* March 15, 1870.

[84] *The Inland Empire,* March 20, 1870.

[85-87] *White Pine News,* April 4, 6, 29, 1869, January 29, 1876.

Health

[88] Henry Eno to William Eno, August 21, 1869. The Henry Eno Papers. Henry Eno, a native of Dutchess County, New York, came to California with the gold rush of 1849, at the age of 51, and spent the next twenty years of his life in the frontier communities of that state and adjoining Nevada. He worked as prospector, judge, and editor, served as a vigilante and eventually became a drifter. Letters written to his brother William from Hamilton, White Pine Mining District, provide the only extant correspondence from a participant in the rush of 1868-69 that the author has been able to locate in major library depositories. They are from Yale University Library.

[89] McDonald, "White Pine County," *loc. cit.,* p. 1044.

[90-91] *White Pine News,* January 23, March 27, April 13, 1869.

[92] April 10, 1869.

[93] *White Pine News,* April 22, 1869.

[94] *The Inland Empire,* April 21, 1869.

[95] *White Pine News,* August 3, 1869.

[96] *The Inland Empire,* August 3, 1869.

[97-98] *White Pine News,* April 5, August 3, 1869.

[99] *The Inland Empire,* August 26, 1869.

[100-101] *White Pine News,* February 20, April 23, 1869.

[102] April 23, 1869.

[103-104] *The Inland Empire,* August 3, 1869.

Newspapers

[105] Angel, *op. cit.,* p. 330.

[106] "Introductory," *White Pine News,* December 26, 1868.

[107] January 30, 1869. Reprint of an article in the Virginia City *Enterprise,* January 21, 1869.

[108] Angel, *op. cit.,* p. 330.

[109] March 27, 1869.

[110] March 31, 1869.

[111] April 3, 1869.

[112-113] April 9, 13, 1869.

[114] Quoted in *The Inland Empire,* April 21, 1869.

[115-116] May 4, 18, 1869.

[117] Quoted in *The Inland Empire,* May 25, 1869.

[118] April 29, 1869.

[119] April 4, 1869.

[120] Angel, *op. cit.,* p. 331; *White Pine Evening Telegram,* June 2, 1869.

[121] July 12, 1869.

[122] *White Pine News,* August 9, 1869; *The Inland Empire,* August 25, 1869.

[123] August 13, 1869.

[124] August 27, 1869.

[125] August 12, 1869.

Lawlessness

[126] Henry Eno to William Eno, June 14 and August 21, 1869. The Henry Eno Papers.

[127] *Alta California,* February 8, 1869.

[128] Henry Eno to William Eno, August 21, 1869. The Henry Eno Papers.

[129] *White Pine News,* April 9, 1869.

[130] *The Inland Empire*, June 13, 1869.
[131-134] *White Pine News*, April 21, May 6, 24, July 7, 1869.
[135] *Alta California*, February 8, 1869.
[136] *White Pine News*, March 6, 1869.
[137] *The Inland Empire*, May 19, 1869.
[138-139] *White Pine Evening Telegram*, June 16, July 21, 1869.
[140] *White Pine News*, April 14, 1869.
[141] *Ibid.*, April 9, 1869; *The Inland Empire*, April 20, 1869.
[142] March 30, 1869.
[143-144] *White Pine News*, February 27, June 7, 1869.
[145] *The Inland Empire*, October 17, 1869.
[146] *White Pine News*, February 25, 1870.
[147] *The Inland Empire*, June 24, 1869.
[148-150] *White Pine News*, April 12, May 6, 12, 19, June 4, 1869.
[151] *Ibid.*, May 5, 7, 1869; *The Inland Empire*, May 5, 18, 1869.
[152-153] *The Inland Empire*, May 1, 1869, March 20, 1870.
[154-158] *White Pine News*, March 20, April 16, August 10, October 2, 9, 1869.
[159-161] *The Inland Empire*, April 30, July 18, October 2, 1869.
[162] *White Pine News*, October 9, 1869.
[163] *Ibid.*, June 11, 1869; *The Inland Empire*, June 11, 1869.
[164] *White Pine News*, July 30, 1869; *The Inland Empire*, July 30, 1869.
[165-169] *White Pine News*, April 16, May 24, 25, 1869, January 8, May 25, 1870.
[170-173] *The Inland Empire*, May 9, June 19, 20, September 11, 17, 1869.
[174-175] *White Pine News*, May 9, 12, 1869, March 9, 1870.
[176-178] *The Inland Empire*, May 28, August 13, 1869, March 11, 1870.
[179] *Ibid.*, March 13, 17, 1870; *White Pine News*, March 17, 1870.

Recreation

[180-181] *White Pine News*, April 9, July 27, 1869.
[182-183] *The Inland Empire*, May 2, June 16, 1869.
[184] *Ibid.*, May 9, 1869; *White Pine News*, May 5, 1869.
[185] *The Inland Empire*, August 5, 1869; *White Pine News*, August 6, 1869.
[186-188] *The Inland Empire*, May 18, September 2, 19, 1869.
[189-194] *White Pine News*, February 20, April 9, 17, 19, 1869, April 20, May 13, 14, June 16, 20, 1870.
[195] *The Inland Empire*, June 19, 1869.
[196] *Ibid.*, August 17, 1869; *White Pine News*, August 16, 1869.
[197] *White Pine Evening Telegram*, June 21, July 17, 28, 1869.
[198] *White Pine News*, May 19, 1870.
[199] *White Pine Evening Telegram*, June 10, 1869.
[200] *The Inland Empire*, July 4, 1869; *White Pine News*, July 8, 1869; *White Pine Evening Telegram*, July 10, 1869.
[201-206] *The Inland Empire*, May 20, July 16, October 2, November 5, 24, 1869, November 5, 1870.
[207-210] *White Pine News*, June 9, 1869, February 6, 23, March 23, 1870.
[211] *Ibid.*, April 20, 1870; *The Inland Empire*, April 19, 1870.
[212] *White Pine News*, May 24, 1869.
[213] *White Pine Evening Telegram*, June 23, 1869.
[214-217] *The Inland Empire*, April 8, 9, 13, June 25, September 24, 1869.
[218] *White Pine News*, April 27, 1869.
[219] *The Inland Empire*, December 29, 1869.
[220] *White Pine News*, March 20, 1869.
[221] *Ibid.*, March 17, 1870; *The Inland Empire*, March 17, 1870.
[222] *White Pine Evening Telegram*, July 6, 1869; *The Inland Empire*, July 7, 1869.
[223-224] *White Pine News*, February 23, June 16, 1870.
[225] *The Inland Empire*, November 20, 1869.
[226-227] *White Pine News*, November 20, December 18, 1869.
[228] *The Inland Empire*, December 23, 25, 1869.
[229-230] *White Pine News*, April 27, June 28, 1869.
[231] *The Inland Empire*, August 12, 1869.
[232] *Ibid.*, August 17, 1869; *White Pine News*, August 17, 1869.

[233-234] *White Pine News,* September 7, 9, 1869.
[235] Henry Eno to William Eno, August 21, 1869. The Henry Eno Papers.
[236] *The Inland Empire,* April 23, 1869; *White Pine News,* November 20, 1869.
[237] *The Inland Empire,* August 6, 1869.

The Theater

[238] *The Inland Empire,* April 2, 1869.
[239-243] *White Pine News,* April 12, 13, 14, 1869.
[244-245] *The Inland Empire,* March 30, April 9, 1869.
[246] *White Pine News,* April 29, 1869.
[247-251] *The Inland Empire,* March 30, May 5, 19, 21, 27, 1869.
[252] *White Pine News,* May 14, 1869.
[253] *The Inland Empire,* May 23, 1869.
[254-255] *White Pine Evening Telegram,* June 4, 5, 1869.
[256] *The Inland Empire,* June 4, 1869.
[257] *White Pine Evening Telegram,* June 9, 1869.
[258-259] *White Pine News,* June 8, 12, 1869.
[260-262] *White Pine Evening Telegram,* June 14, 18, July 23, 1869.
[263-268] *The Inland Empire,* June 15, 26, 30, July 18, 29, August 3, September 7, 1869.
[269-271] *White Pine News,* April 22, May 7, 10, 27, 1870, April 26, May 21, 1872.

The Church, the School, and Humanitarianism

[272] *White Pine Evening Telegram,* July 17,1869.
[273] *White Pine News,* February 20, 27, 1869.
[274] *Reese River Reveille* quoted in *White Pine News,* June 22, 1869.
[275-278] *The Inland Empire,* April 13, 15, 24, 1869, February 8, 1870.
[279] *White Pine News,* June 25, 1869.
[280-281] *White Pine Evening Telegram,* June 26, 30, 1869.
[282] *White Pine News,* June 28, 1869.
[283-289] *The Inland Empire,* July 1, 13, August 11, 27, September 22, November 30, 1869, February 17, 1870.
[290-292] *White Pine News,* April 13, 23, 1870, January 29, 1876.
[293-296] *The Inland Empire,* July 8, August 28, December 4, 31, 1869, February 26, 1870.
[297-301] *White Pine News,* March 7, April 15, 1869, March 5, April 7, May 11, 1870.
[302] *Ibid.,* March 30, April 1, 2, 1870; *The Inland Empire,* March 30, April 1, 1870.
[303] *White Pine News,* April 7, 1870.
[304] *The Inland Empire,* March 29, 1869.
[305-308] *White Pine News,* August 11, 1869, June 18, July 16, 21, 1870.
[309] Information obtained by the author at Hamilton, Nevada, August, 1955.
[310] *White Pine News,* July 28, 1870.

Local Government

[311-314] *White Pine News,* December 26, 1868, January 2, 30, 1869.
[315-316] "An Act to create the County of White Pine, and provide for its organization, March 2, 1869, *"Statutes of the State of Nevada,* Fourth Session, 1869, pp. 108-110.
[317-318] *White Pine News,* March 6, 20, 1869.
[319-321] *The Inland Empire,* June 10, 1868, April 13, 1869.
[322] June 11, 1869.
[323-324] *Statutes of the State of Nevada,* Fourth Session, 1869, pp. 133-134, 153-157.
[325] *White Pine News,* March 20, 1869.
[326] *Statutes of the State of Nevada,* Fourth Session, 1869, pp. 122-127.
[327-329] *The Inland Empire,* April 8, 16, 27, 1869.
[330] *White Pine News,* May 12, 1869.
[331] *The Inland Empire,* May 21, 1869.
[332] *White Pine News,* May 28, 1869.

[333] *The Inland Empire*, May 29, 1869.
[334-336] *White Pine News*, May 27, June 5, 8, 1869.
[337] *The Inland Empire*, June 9, 1869.
[338] *White Pine Evening Telegram*, June 5, 8, 1869.
[339-340] *The Inland Empire*, July 29, August 4, 15, 1869.
[341] *White Pine Evening Telegram*, August 17, 1869.
[342-343] *The Inland Empire*, August 27, December 23, 1869.
[344] Angel, *op. cit.*, pp. 659.
[345] *The Inland Empire*, March 17, 1870.
[346-347] *White Pine News*, January 22, March 17, 1870.
[348] B. F. Miller, "Nevada in the Making: Being Pioneer Stories of White Pine County and Elsewhere," *Nevada State Historical Society Papers, 1923-1924*, p. 276.
[349] *White Pine News*, February 7, 1870.
[350] *The Inland Empire*, February 20, 1870.
[351] February 21, 1870.
[352] *The Inland Empire*, February 22, 24, 1870.

Partisan Politics

[353] *Mining and Scientific Press, XIX* (July 10, 1869), p. 18.
[354-355] *The Inland Empire*, July 11, 17, 1869.
[356] *White Pine Evening Telegram*, July 19, 1869.
[357] *White Pine News*, July 19, 1869.
[358] *The Inland Empire*, July 22, 1869.
[359] *White Pine Evening Telegram*, July 23, 1869.
[360-361] *White Pine News*, July 22, 24, 1869.
[362] *The Inland Empire*, March 22, April 10, 1870.
[363-373] *White Pine News*, March 22, May 14, 16, 19, 20, 25, 30, June 6, 7, 8, 1870.
[374] Angel, *op. cit.*, p. 330.
[375-381] *The Inland Empire*, October 14, 18, 19, 26, November 2, 3, 8, 9, 1870.
[382] James G. Scrugham, *Nevada, A Narrative of the Conquest of a Frontier Land*, I, p. 271.
[383] Angel, *op. cit.*, p. 331.

Property and Prosperity

[384] Evans, *op. cit.*, p. 23.
[385] February 8, 1869.
[386] *White Pine News*, February 13, 1869.
[387] Henry Eno to William Eno, June 14, 1869. The Henry Eno Papers.
[388-389] *White Pine News*, February 13, 20, 1869.
[390] *The Inland Empire*, May 20, 1869.
[391-392] *White Pine News*, May 6, 12, 1869.
[393] April 14, 1869.
[394-395] *The Inland Empire*, June 19, July 3, 1869.
[396] *White Pine News*, February 13, 1869.
[397] *Mining and Scientific Press, XX* (June 11, 1870), p. 394.
[398-399] *White Pine News*, March 20, April 15, 1869.
[400] *The Inland Empire*, July 27, 1869.
[401] Henry Eno to William Eno, August 21, 1869. The Henry Eno Papers.
[402] *White Pine News*, June 11, 1869.
[403] *The Inland Empire*, September 19, 1869.
[404-405] *Mining and Scientific Press, XIX* (August 21, 1869), p. 114; XV (June 11, 1870), p. 394.
[406] August 9, 1869.
[407] *The Inland Empire*, December 22, 1869.
[408-409] Evans, *op. cit.*, pp. 18-19.
[410-412] *White Pine News*, March 20, April 6, August 12, 1869.
[413] *The Inland Empire*, April 30, 1869.
[414] Campton, "Experiences of a Nevada Pioneer," *loc. cit.*
[415] Angel, *op. cit.*, p. 663.

[416-417] *The Inland Empire,* May 15, 18, 1869.
[418-419] *White Pine News,* April 16, June 24, 1869.
[420] *White Pine Evening Telegram,* June 3, 1869.
[421] San Francisco *Bulletin,* March 8, 1869.
[422] Angel, *op. cit.,* p. 650.
[423] *Report of the Nevada State Mineralogist, 1869–1870,* pp. 70, 74-75.
[424] Hague, "Mining Industry," *loc. cit.,* p. 439; *White Pine News,* January 2, 1869.
[425-428] *White Pine News,* March 20, April 10, 14, 1869.
[429] Henry Eno to William Eno, August 8, 1869. The Henry Eno Papers.
[430-431] *The Inland Empire,* March 31, April 30, 1869.
[432] *Mining and Scientific Press,* XVIII (April 3, 1869), p. 221.
[433-434] *White Pine News,* January 16, April 15, 1869.
[435] *The Inland Empire,* March 27, 1869.
[436] Hague, Mining Industry, *loc. cit.,* p. 437.
[437-438] *White Pine News,* February 27, May 8, 1869.
[439] Henry Eno to William Eno, August 21, 1869. The Henry Eno Papers.
[440] *White Pine News,* June 15, 1869.
[441] Hague, "Mining Industry," *loc. cit.,* p. 437.
[442] *White Pine News,* October 10, 1869.
[443] *Report of the Nevada State Mineralogist, 1869–1870,* p. 77.
[444] Lorsh, "Mining at Hamilton, Nevada," *loc. cit.,* p. 521.
[445] *White Pine Evening Telegram,* July 21, 1869.
[446] Angel, *op. cit.,* p. 663.
[447] *White Pine News,* June 2, 1869.
[448] Angel, *op. cit.,* p. 663.
[449] Lorsh, "Mining at Hamilton, Nevada," *loc. cit.,* p. 521.
[450] *Mining and Scientific Press,* XIX (July 10, 1869), p. 18.
[451] *The Inland Empire,* June 4, 1869.

Chapter III: DECLINE

[1] *Alta California,* March 11, 1869.
[2-3] *The Inland Empire,* April 18, August 24, 1869.
[4-5] *White Pine News,* April 20, 22, 1869.

The Struggle of Labor

[6-7] *White Pine News,* October 23, 1869, February 9, 1870.
[8] *The Inland Empire,* June 5, 1869.
[9] April 12, 1869.
[10] July 13, 1869.
[11] July 14, 1869.
[12-15] *White Pine News,* July 12, 13, 14, 1869.
[16-17] *White Pine Evening Telegram,* July 14, 17, 1869.
[18-19] *The Inland Empire,* July 16, 27, 1869.
[20] *White Pine Evening Telegram,* July 27, 1869.
[21] July 28, 1869.
[22] July 28, 1869.
[23] July 28, 1869.
[24-25] *The Inland Empire,* July 29, 1869.
[26] *White Pine News,* July 30, 1869.
[27] *White Pine Evening Telegram,* July 31, 1869; *The Inland Empire,* July 31, 1869.
[28] *White Pine News,* August 2, 1869.
[29] *White Pine Evening Telegram,* August 3, 1869.
[30-31] *White Pine News,* August 5, 1869.
[32] August 5, 1869.
[33] August 5, 1869.
[34] *White Pine News,* August 5, 1869.
[35] *White Pine Evening Telegram,* August 6, 1869.

[36-37] *White Pine News,* August 6, 7, 1869.
[38] *The Inland Empire,* August 8, 1869.
[39-40] August 9, 1869.

The Problem of Water and the Peril of Fire

[41] Evans, *op. cit.,* p. 37.
[42] *White Pine News,* December 26, 1868.
[43] Evans, *op. cit.,* p. 37.
[44-45] *White Pine News,* April 7, May 7, 1869.
[46] *White Pine Evening Telegram,* August 2, 1869.
[47-49] *White Pine News,* February 13, 27, May 7, 1869.
[50] *The Inland Empire,* February 12, 1870.
[51] April 9, 1869.
[52] *The Inland Empire,* May 13, 1869.
[53-54] *White Pine News,* September 28, 1869, August 4, 1870.
[55-56] *The Inland Empire,* July 13, November 12, 1869.
[57] Angel, *op. cit.,* p. 660.
[58] Hague, "Mining Industry," *loc. cit.,* p. 442.
[59-61] *The Inland Empire,* April 15, June 5, July 9, 1869.
[62] *White Pine News,* August 14, 1869.
[63-65] *The Inland Empire,* September 2, 5, 15, 1869.
[66] *White Pine News,* September 16, 1869.
[67] September 24, 1869.
[68-69] *White Pine News,* August 27, September 23, 1869.
[70] *The Inland Empire,* October 10, 1869.
[71-72] *White Pine News,* October 16, 29, 1869.
[73-74] *The Inland Empire,* October 26, November 11, 1869.
[75-76] *White Pine News,* February 6, March 17, 1870.
[77]March 18, 19, 1870.
[78-79] *White Fine News, March* 19, 21, 1870.
[80-81] *The Inland Empire,* March 23, 31, 1870.
[52] *The Inland Empire,* April 5, 1870; *White Pine News,* April 5, 1870.
[83-86] *White Pine News,* April 11, 18, 19, May 9, June 29, 1870.
[87] Angel, *op. cit.,* p. 660.
[88] *White Pine News,* June 27, 1873, quoted in *ibid.*
[89] Angel, *op. cit.,* pp. 660–661.

The Search for Capital

[90] *The Inland Empire,* June 4, August 6, 1869; *White Pine Evening Telegram,* August 6, 1869.
[91] *Mining and Scientific Press,* XIX (August 21, 1869), p. 114; *White Pine News,* August 14, 1869.
[92] *White Pine News,* September 14, 1869.
[93] *The Inland Empire,* September 22, 1869.
[94] *White Pine News,* January 15, 1870.
[95] Hague, Mining Industry, *loc. cit.,* p. 439.
[96] February 17, 1870.
[97] *The Inland Empire,* June 4, 1869.
[98] *White Pine News,* August 21, 1869.
[99-100] *Chicago Tribune,* October 19, 1869, quoted in *The Inland Empire,* October 26, 1869.
[101] November 4, 1869.
[102] Henry Eno to William Eno, August 8, August 21, 1869. The Henry Eno Papers.
[103] *White Pine News,* January 22, 1870.
[104] Bancroft, *op. cit.,* pp. 278, 279.
[105] Lincoln, *op. cit.,* p. 257; McDonald, "White Pine County," *loc. cit.,* p. 1044.
[106] *The Inland Empire,* April 8, 1870.

The Turning Point

[107] *The Inland Empire,* October 7, 8, 1869; *White Pine News,* October 9, 1869.
[108] *The Inland Empire,* October 8, 1869.
[109-111] *White Pine News,* October 16, 26, November 6, 1869.
[112-113] *The Inland Empire,* December 12, 17, 1869.
[114] *White Pine News,* January 1, 1870.
[115-118] *The Inland Empire,* December 22, 1869, February 1, 10, 12, 1870.
[119] August 28, 1869.
[120] *White Pine Evening Telegram,* August 11, 1869; *The Inland Empire,* August 11, 1869.
[121] August 21, 1869.
[122] October 2, 1869.
[123-124] October 21, November 4, 28, 1869.
[125] January 8, 1870.
[126-127] *White Pine News,* March 18, April 4, 13, 1870.
[128] *Ibid.,* April 25, 28, May 10, 1870; Angel, *op. cit.,* p. 331.
[129-130] *White Pine News,* May 18, June 15, 1870.
[131] Angel, *op. cit.,* p. 331.
[132] July 22, 1870.
[133-135] *White Pine News,* March 29, May 12, July 16, 1870.
[136] *White Pine News,* March 27, 1869; *The Inland Empire,* April 28, 1869.
[137] *Statutes of the State of Nevada,* Fourth Session, 1869, pp. 160–162.
[138] *White Pine News,* March 31, 1870.
[139] *The Inland Empire,* April 3, 1870.
[140-146] *White Pine News,* May 10, 14, 19, 30, June 3, 7, 10, 1870.
[147-148] *Statutes of the State of Nevada,* Fifth Session, 1871, pp. 67–70; Sixth Session, 1873, pp. 219–222.
[149] Angel, *op. cit.,* pp. 283–285, 289.

Chapter IV: RALLY AND COLLAPSE

[1] Clark C. Spence, *British Investments and the American Mining Frontier, 1860–1901,* p. 62. Prospectus, Papers of the Washoe United Consolidated Gold and Silver Mining Company, Limited, Companies Registration Office, Bancroft Library Research Program for the Collection of Western Americana in Great Britain. Hereafter cited, CRO, BLRP.
[2] London *Mining Journal,* LXVII (October 30, 1897), p. 1280.
[3] Prospectuses of Austin Consolidated Silver Mines Company, Limited, Reese River Silver Mining Company, Limited, and Lander City Silver Mining Company, Limited, CRO, BLRP.
[4] *American Journal of Mining,* V (February 15, 1868), p. 104–105.
[5] *White Pine Evening Telegram,* June 3, 1869.
[6] *White Pine News,* June 12, 1869; *White Pine Evening Telegram,* June 14, 1869.
[7] *White Pine News,* February 6 and 7, 1870.
[8] Spence, *op. cit.,* p. 5, indicates that eleven of the sixteen Anglo-American mining companies floated between 1864 and 1870, whose records he has found, were located in Nevada. Between 1870–1873, the number of British enterprises in Nevada rapidly increased.
[9] Memorandum of Association, Papers of the Eberhardt and Aurora Mining Company, Limited, CRO, BLRP.
[10] London *Mining Journal,* XL (March 26, 1870), p. 248.
[11] Summary of Capital and Shares, July, 1870, Papers of the Eberhardt and Aurora Mining Company, Limited, CRO, BLRP.
[12-14] *White Pine News,* April 1, 4, 11, 19, 20, June 8, 1870.
[15] *Mining and Scientific Press,* XX (April 16, 1870), p. 250.
[16] London *Mining World,* June 25, 1870, p. 531.
[17] London *Mining Journal,* XL (July 26, 1870), p. 626.
[18-20] *White Pine News,* June 15, 17, July 11, 1870.
[21] Property Deeds, William Miles Read Papers, The Bancroft Library.
[22] *Mining and Scientific Press,* XXIII (September 30, 1871), p. 195.

[23] Document on Records and Accounts, Eberhardt and Aurora Mining Company, Limited, Read Papers.
[24] Thomas Phillpotts Correspondence, Read Papers.
[25] *White Pine News,* July 27, 30, 1870.
[26] *Mining and Scientific Press,* XXI (August 20, 1870), p. 132.

British Contributions

[27] Phillpotts to Lighter, September 11, 1870, Read Papers.
[28-30] George Attwood to Phillpotts, August 14, September 4, 5, 8, 9, 1870, Read Papers.
[31] London *Mining Journal,* XL (November 19, 1870), p. 984.
[32] *The Inland Empire,* October 8, 11, 15, November 2, 4, 5, 6, 1870.
[33] London *Mining Journal,* XL (October 8, 1870), p. 841.
[34] *White Pine News,* April 21, 1870.
[35] London *Mining Journal,* XL (November 19, 1870), p. 984.
[36] Memorandum of Association, Selected Papers of the South Aurora Silver Mining Company, Limited, CRO, BLRP.
[37] Articles of Association, Selected Papers of the South Aurora Silver Mining Company, Limited, CRO, BLRP.
[38] London *Mining Journal,* XL (October 29, 1870), p. 908.
[39] Summary of Capital and Shares, February 1, 1871, Selected Papers of the South Aurora Silver Mining Company, Limited, CRO, BLRP.
[40] *Mining and Scientific Press,* XXI (September 24, 1870), p. 218; XXII (January 28, 1871), p. 60.
[41] Agreement of January 16, 1871, between Edward Applegarth and the Eberhardt and Aurora Mining Company, Limited, CRO, BLRP. Copy of Contract in Read Papers.
[42-43] London *Mining Journal,* XLI (March 11, 1871), p. 211; (April 1, 1871), p. 264; (April 8, 1871), p. 284.
[44] *White Pine News,* April 3, 1871.
[45] London *Mining Journal,* XLI, (May 20, 1871), pp. 428, 439; (May 27, 1871), p. 449.
[46] *Mining and Scientific Press,* XXV (August 10, 1872), p. 82.
[47] XII (July 11, 1871), p. 25.
[48] August 16, 1869.
[49] Tramway records, including contracts, blueprints, specifications, and correspondence, Read Papers.
[50] *Mining and Scientific Press,* XXIII (October 7, 1871), p. 211; XXV (August 3, 1872), p. 66.
[51] Manuscript, Read Papers.
[52] *Mining and Scientific Press,* XXIII (October 7, 1871), p. 211; Correspondence, Read Papers.
[53-56] London *Mining Journal,* XLI (July 15, 1871), p. 604; (September 9, 1871), pp. 785, 800; (September 30, 1871), p. 852; (October 7, 1871), p. 872; (October 14, 1871), p. 911; (October 21, 1871), pp. 917, 932; (November 4, 1871), p. 963; (November 25, 1871), pp. 1035–1036.
[57-58] Correspondence, reports, deeds, and specification of the White Pine Water Works Company, Read Papers.
[59] *White Pine News,* March 30, 1872.
[60] C. T. Bulkley, civil engineer, to F. H. Benjamin, superintendent of the White Pine Water Works, July 5, 1871.
[61] Thomas Wren to Phillpotts, June 13, 1871, Read Papers.
[62] *White Pine News,* April 2, 5, 1872.
[63] Deed from Blasdel to Phillpotts, October 14, 1872.
[64] Phillpotts to Attwood, June 16, 1871; Attwood to Phillpotts, June 20, 1871, Read Papers.
[65] Statement of Expenditures, September, 1870 to October, 1871, Read Papers.
[66-67] London *Mining Journal,* XLI (December 9, 1871), p. 1098; (December 16, 1871), p. 1123.
[68] Bullion Ledger, Eberhardt and Aurora Mining Company, Limited, Read Papers.

[69] *White Pine News,* April 16, May 16, 1872.
[70] Statement of Expenditures, October, 1871 to September 1872, Read Papers.
[71] *Mining and Scientific Press,* XXIII (October 7, 1871), p. 211.
[72] *White Pine News,* May 17, 1872.
[73-75] London *Mining Journal,* XLII (September 28, 1872), p. 928; (October 19, 1872), pp. 984, 1005–1006; (October 26, 1872), p. 1029.
[76] Memorandum of Instructions to Frank Drake, February 28, 1873, Read Papers.
[77] Kimber and Ellis, Solicitors, to the Eberhardt and Aurora Mining Company, March 31, 1873, Read Papers.
[78] London *Mining World,* August 16, 1873, p. 364.
[79] Spence, *op. cit.,* pp. 116-117, 166, 170.
[80-81] London *Mining Journal,* XLII (February 3, 1872), p. 95; (November 9, 1872), pp. 1078–1079; (November 16, 1872), p. 1092; (November 23, 1872), p. 1127.
[82] *Ibid.,* (December 7, 1872), p. 1168; XLIII (January 18, 1873), p. 59; Special Resolution of December 19, 1872, Confirmed, January 16, 1873, Selected Papers of the South Aurora Silver Mining Company, Limited, CRO, BLRP.
[83] *Report of the Nevada State Mineralogist, 1871-1872,* p. 143.

British Determination and Adversity

[84] *Mining and Scientific Press,* XXVIII (January 31, 1874), p. 74.
[85] Ibid., XXVII (October 4, 1873), p. 216; Record of Ore Assays, Eberhardt and Aurora Mining Company, Limited, Read Papers.
[86] *White Pine News,* quoted in *Mining and Scientific Press,* XXVIII (February 7, 1874), p. 86.
[87] *Report of the Nevada State Mineralogist, 1873–1874,* pp. 84–85.
[88] Miller, "Nevada in the Making: Being Pioneer Stories of White Pine County and Elsewhere," *loc. cit.,* p. 283.
[89] *Mining and Scientific Press,* XXX (January 23, 1875), p. 50.
[90] *Engineering and Mining Journal,* XVII (January 31, 1874), p. 68.
[91] London *Mining Journal,* XLIV (August 29, 1874), p. 949; London *Mining World,* August 29, 1874, clipping in Read Papers.
[92-98] London *Mining Journal,* XLIV (January 31, 1874), p. 131; (May 9, 1874), p. 511; (September 7, 1874), p. 1033; (November 28, 1874), p. 1313; XLV (June 5, 1875), p. 623; (June 26, 1875), pp. 705, 706; (December 18, 1875), pp. 1405–1406.
[99] Contract of December 24, 1875, Selected Papers of the Eberhardt and Aurora Mining Company, Limited, CRO, BLRP.
[100] *White Pine News,* December 11, 1875.
[101] XXXI (December 18, 1875), p. 385.
[102] *White Pine News,* December 4, 1875.
[103] *Mining and Scientific Press,* XXXI (December 18, 1875), p. 385.
[104] *Ibid.,* XXXII (March 4, 1876), p. 146; *White Pine News,* February 19, 1876. The assessor's figures do not coincide exactly with those of the state mineralogist but the trend is the same.
[105] London *Mining Journal,* XLV (December 18, 1875), pp. 1405–1406.
[106-108] *White Pine News,* February 12, 26, March 25, 1876.
[109] London *Mining Journal,* XLVI (April 29, 1876), p. 482; London *Mining World,* April 29, 1876, quoted in *White Pine News,* May 20, 1876.
[110] London *Mining Journal,* XLVI (January 15, 1876), p. 72.
[111-118] *White Pine News,* December 4, 1875, March 4, April 1, 15, 29, May 6, 20, June 3, August 5, 26, September 9, October 7, December 2, 16, 30, 1876, January 13, 1877.
[119] London *Mining World,* July 1, 1876, quoted in *White Pine News,* July 22, 1876.
[120] London *Mining Journal,* XLVII (January 6, 1877), p. 22.
[121] *Ibid.,* (February 10, 1877), p. 140; London *Mining World,* February 17, 1877, quoted in *White Pine News,* March 10, 1877.
[122] *White Pine News,* March 24, April 14, 1877.

[123-124] London *Mining Journal,* XLVII (May 19, 1877), p. 547; XLVIII (May 11, 1878), p. 523.

[125] Agreement of October 29, 1878, Papers of the Eberhardt and Aurora Mining Company, Limited, CRO, BLRP.

[126-127] London *Mining Journal,* XLVIII (June 1, 1878), p. 588; (June 22, 1878), p. 695; (October 12, 1878), pp. 1137–1138.

[128] Papers of the Consolidated Mining Company, Limited, CRO, BLRP.

[129] London *Mining Journal,* XLIX (May 31, 1879), p. 558.

[130] *Report of the Nevada State Mineralogist, 1877-1878,* p. 155.

[131] London *Mining Journal,* L (February 21, 1880), pp. 219–220; (February 28, 1880), p. 248.

[132-133] *White Pine News,* May 6, September 2, 1880.

[134] Quoted in *ibid.,* September 9, 1880.

[135] London *Mining Journal,* L (December 25, 1880), p. 1495; Special Resolutions passed December 21, 1880 and confirmed January 5, 1881, Papers of the Eberhardt and Aurora Mining Company, Limited. Each holder of a £100 debenture was to receive 100 fully-paid shares in the new company and 100 fully-paid preference shares in addition. The latter were to draw ten per cent preferential dividend.

[136] London *Mining World,* January 8, 1881, quoted in *White Pine News,* January 29, 1881; *ibid.,* January 15, 1881.

[137] *White Pine News,* September 3, October 29, 1881. The Eureka *Sentinel* was the greatest offender in starting rumors that other newspapers reprinted.

[138] London *Mining Journal,* LII (June 3, 1882), p. 680.

[139] Miller, "Nevada in the Making: Being Pioneer Stories of White Pine County and Elsewhere," *loc. cit.,* p. 288.

[140-141] London *Mining Journal,* LIII (June 9, 1883), p. 663; LIV (February 9, 1884), p. 152; (March 1, 1884), p. 244; (March 22, 1884), p. 335.

[142] Report on Mining Properties of the Eberhardt Company, Limited, April 14, 1885; by Thomas Price, Mining Engineer, Read Papers.

Abandonment

[143-144] *White Pine News,* quoted in *Mining and Scientific Press,* XXVIII (April 11, 1874), p. 230.

[145] *White Pine News,* July 13, 1870.

[146-149] *Statutes of the State of Nevada,* Fifth Session, 1871, pp. 48–49, 75, 95–102; Sixth Session, 1873, p. 88.

[150-151] *White Pine News,* quoted in *Mining and Scientific Press,* XXVI (June 28, 1873), p. 410; XXVII (November 29, 1873), p. 342.

[152-155] *Mining and Scientific Press,* XXVIII (February 7, 1874), p. 86; (February 28, 1874), p. 134; (March 14, 1874), p. 170; (April 11, 1874), p. 230.

[156] *Report of the Nevada State Mineralogist, 1873-1874,* pp. 84–85.

[157] *Statutes of the State of Nevada,* Seventh Session, 1875, p. 97.

[158-161] *White Pine News,* November 27, December 25, 1875, January 15, October 28, 1876, February 17, 1877.

[162] *Mining and Scientific Press,* XXXI, (November 27, 1875), p. 344; *White Pine News,* November 13, 1875.

[163] *White Pine News,* December 23, 1876, January 27, February 17, 1877.

[164] Angel, *op. cit.,* p. 330.

[165] *White Pine News,* August 12, October 28, 1880.

[166-167] Angel, *op. cit.,* pp. 662–663.

[168-178] *White Pine News,* November 18, December 16, 1880, January 8, July 16, 1881, January 21, February 4, March 25, April 1, 22, 29, June 3, 1882.

[179-180] Miller, "Nevada in the Making: Being Pioneer Stories of White Pine County and Elsewhere," *loc. cit.,* pp. 274-275, 278-279, 280-282.

[181] Report of Frank Drake to the Eberhardt Company, Limited, May 18, 1885, Read Papers.

[182] London *Mining Journal,* LV (August 8, 1885), p. 889; *Engineering and Mining Journal,* XL (August 29, 1885), p. 152.

[183] Papers of the Eberhardt and Moniter Company, Limited, CRO, BLRP; London *Mining Journal,* LV (October 17, 1885), p. 1171.

[184] *Engineering and Mining Journal,* XLIII (January 1, 1887), p. 8.

[185-186] London *Mining Journal,* LV (November 21, 1885), pp. 1314–1315; LVI (January 23, 1886), p. 98; (November 27, 1886), p. 1365; LVII (August 13, 1887), pp. 991–992; LVIII (March 10, 1888), p. 261; LIX (April 20, 1889), p. 448. Papers of the Consolidated Mining Company, Limited, CRO, BLRP.

[187] Statement of Mrs. Dora Hoover, Read Papers. Mrs. Hoover is the daughter of William Miles Read.

[188] Statement of Frank Drake, December 26, 1896, Read Papers.

[189] London *Mining Journal,* LVIII (March 24, 1888), p. 321.

[190] Indentures, Read Papers.

[191] Papers of the New Eberhardt Company, Limited, CRO, BLRP: *Engineering and Mining Journal,* XLV (May 5, 1888), p. 330; XLVI (August 18, 1888), p. 136; London *Mining Journal,* LVIII (April 7, 1888), p. 382, (June 30, 1888), pp. 733–734; (July 21, 1888), p. 819; (September 1, 1888), p. 987.

[192] *The Financial World,* IV (December 8, 1888), Clipping in Read Papers.

[193] London *Mining Journal,* LIX (July 6, 1889), p. 765; LX (May 24, 1890), p. 591; *Engineering and Mining Journal,* LI (June 27, 1891), p. 751.

[194] Statement of Mrs. Dora Hoover, Read Papers.

[195] Read-Oxenford Correspondence, 1891-1893. Printed reports of the New Eberhardt Company, 1889, 1891, 1892. Separates in Read Papers.

[196] *Engineering and Mining Journal,* LV (June 17, 1893), pp. 567, 569.

[197] Power of Attorney to William Read, February 1, 1893; Deeds and indentures, September 8, 1896, Read Papers.

[198] The Thistle Consolidated Mines, Limited, to Read, January 13, 1896, Read Papers.

[199] Report on Treasure Hill Properties by K. Freitig, 1916, Read Papers.

[200] Read to Oxenford, September 25, 1900, Read Papers.

BIBLIOGRAPHY

Manuscript Collections

Bancroft Library Research Program for the Collection of Western Americana in Great Britain (University of California, Berkeley):

Records from the Companies Registration Office, Bush House, London:

Microfilmed public records including Memoranda of Association, Articles of Association, Annual Summaries of Capital and Shares, Periodic Lists of Stockholders, Winding-Up Records, etc.

Eberhardt and Aurora Mining Company, Limited, 1870–1881
The Eberhardt Company, Limited, 1881–1885
Eberhardt and Monitor Company, Limited, 1885–1888
The New Eberhardt Company, Limited, 1888–1893

South Aurora Silver Mining Company, Limited, 1870–1873
The South Aurora Consolidated Company, Limited, 1873–1878
The Consolidated Mining Company, Limited, 1878–1886

Henry Eno Papers, Western Americana Collection, Yale University Library, New Haven.

Three letters written from Hamilton, White Pine County, Nevada during July-August, 1869.

William Miles Read Papers. Bancroft Library, University of California, Berkeley.

Personal papers of the superintendent of Treasure Hill properties of the British. The collection includes eleven maps, a score of photographs, patents to land, original location notices, transactions and specifications for the construction of a water works, a wire tramway, and a stamp mill near Treasure Hill, annual ledgers, payroll sheets, personal and business correspondence.

Newspapers

Ely, Nevada: [Place varies: Treasure City, Hamilton, Cherry Creek, and Ely.]
White Pine News   December 26, 1868–August 7, 1870;
March 30–May 23, 1872;
October 16, 1875–April 14, 1877;
May 6–December 16, 1880;
January 8, 1881–July 29, 1882.

Hamilton, Nevada:
    The Inland Empire   March 27–December 31, 1869;
                                     February 1–April 10, 1870;
                                     October 4–November 9, 1870.

Shermantown, Nevada:
    White Pine Evening Telegram   June 2–August 18, 1869.

San Francisco, California:
    Alta California.   Scattered issues.
    Bulletin.   Scattered issues.

Mining Guides

Cadwallader, General Allen (compiler). *Map and Guide to the White Pine Mines and the Region of Country Adjacent in Eastern Nevada* with complete data to April, 1869, regarding to topography, climate, geological formation, character of ores, vein system of the district, table of altitudes, distances, etc., descriptions of towns, mines, mill sites, and other improvements. San Francisco: H. H. Bancroft and Company, J. Winterburn and Co., Printers, 1869.

Evans, Albert S. *White Pine:* Its Geographical Location, Topography, Geological Formation; Mining Laws; Mineral Resources; Towns; Surroundings; Climate, Population, Altitude, and General Characteristics; Conditions of Society; How to Reach There; What It Costs to Get There and Life There; When to Go There, etc., etc. San Francisco: Alta California Printing House, 1869.

Tagliabue and Barker. *Map of the White Pine Mines and the Regions Adjacent* with an Essay on the Geology and vein system of the District and the character of the Surrounding Country, accompanied by table of Heights, Distances, Bullion Products, etc. San Francisco: Francis and Valentine, Commercial Steam Printing House, 1869.

Periodicals

American Journal of Mining. New York, 1866–1868.

Engineering and Mining Journal. New York, 1870–1893. A continuation of the *American Journal of Mining.*

The Financial World: Mines, Trading Companies, Stock and Shares. London, Vol. IV, 1888.

The Mining Journal. London, Volumes 40–60, 1870–1890.

The Mining World. London, Scattered volumes and issues, 1870–1895.

Mining and Scientific Press. San Francisco, Volumes XVII–XXXII, 1868–1876.

Skinner, Thomas, *The Stock Exchange Year-Book, 1879, 1885, 1890, 1895,* containing a careful digest of information related to the origin, history, and present position of the Joint Stock Companies and Public Securities known to the markets of the United Kingdom. London, Paris, New York: Cassell Petter and Galpin, annual.

Government Documents

Federal:

    A Compendium of the Ninth Census, June 1, 1870. Compiled by Francis A. Walker, Superintendent of Census. 42 Congress, 1 Session, 1872.

    Humphreys, A. A. *Preliminary Report Concerning Explorations and Surveys Principally in Nevada and Arizona, 1871.* [War Department Report]. Washington: Government Printing Office, 1872.

    King, Clarence. *United States Geological Exploration of the Fortieth Parallel.* Volume III. *Mining Industry,* by James D. Hague. Washington: Government Printing Office, 1870.

State:

    Reports of the Nevada State Mineralogist, 1867–1868, 1869–1870, 1871–1872, 1873–1874, 1875–1876, 1877–1878.

Statutes of the State of Nevada, Fourth Session of the Legislature, 1869. Carson City: Henry Mighels, State Printer, 1869.
Statutes of the State of Nevada, Fifth Session of the Legislature, 1871.
Statutes of the State of Nevada, Sixth Session of the Legislature, 1873.
Statutes of the State of Nevada, Seventh Session of the Legislature, 1875.

Articles

Campton, Aaron D., "Experiences of a Nevada Pioneer," *Second Biennial Report of the Nevada Historical Society, 1909–1910* (Carson City, 1911), p. 103.
Elliott, Russell Richard, "The Early History of White Pine County, Nevada, 1865–1871," *Pacific Northwest Quarterly,* XXX (April, 1939), pp. 145-168.
Gracey, Charles, "Early Days in Lincoln County," *First Biennial Report of the Nevada Historical Society, 1907–1908* (Carson City, 1909), pp. 103-105.
Lorsh, W. S., "Mining at Hamilton, Nevada," *Mines and Minerals,* Vol. 29, August, 1908–July, 1909 (Scranton, Pennsylvania: International Textbook Company, 1909), pp. 521-523.
Miller, B. F., "Nevada in the Making: Being Pioneer Stories of White Pine County and Elsewhere," *"Nevada State Historical Society Papers, 1923–1924* (Reno, Nevada, 1924), pp. 272-291.
"White Pine," *Overland Monthly,* III (San Francisco, September, 1869), pp. 209-210.

Books and Pamphlets

Angel, Myron. *History of Nevada: With Illustrations and Biographical Sketches of Its Prominent Men and Pioneers.* Oakland: Thompson and West, 1881.
Baker, James H. and LeRoy R. Hafen. *History of Colorado.* Denver: Linderman Company, 1927. Volume II.
Bancroft, Hubert Howe. *History of Nevada: Colorado, and Wyoming, 1540–1888* [*Works,* Volume XXV]. San Francisco: The History Company, 1890.
Cleland, Robert Glass. *A History of California: The American Period.* New York: The Macmillan Company, 1930.
Davis, Sam P. *History of Nevada.* Reno, Nevada, and Los Angeles, California: The Elms Publishing Company, Inc., 1913.
DeQuille, Dan [William Wright]. *The Big Bonanza.* New York: Alfred A. Knopf, 1947.
Fritz, Percy S. *Colorado, The Centennial State.* New York: Prentice-Hall, Inc., 1941.
Lewis, Oscar. *Silver Kings.* New York: Alfred A. Knopf, 1947.
Lincoln, Francis Church. *Mining Districts and Mineral Resources of Nevada.* Reno: Nevada Newsletter Publishing Company, 1923.
Lord, Eliot. *Comstock Mining and Miners.* [Monographs of the United States Geological Survey, Volume IV.] Washington: Government Printing Office, 1883.
Lyman, George D. *Ralston's Ring.* New York and London: Charles Scribner's Sons, 1937.
Lyman, George D. *The Saga of the Comstock Lode.* New York and London: Charles Scribner's Sons, 1934.
Mack, Effie Mona. *Nevada: A History of the State from the Earliest Times Through the Civil War.* Glendale: The Arthur H. Clark Company, 1936.
Manter, Ethel. *Rocket of the Comstock.* Caldwell, Idaho: The Caxton Printers, Ltd., 1950.
Michelson, Miriam. *The Wonderlode of Silver and Gold.* Boston: The Stratford Company, 1934.
Morrell, W. P. *The Gold Rushes.* New York: The Macmillan Company, 1941.
Paul, Rodman Wilson. *California Gold: The Beginning of Mining in the Far West.* Cambridge: Harvard University Press, 1947.
Powell, John J. *Nevada: The Land of Silver.* San Francisco: Bacon and Company, 1876.

Scrugham, James G. *Nevada: A Narrative of the Conquest of a Frontier Land.* Chicago and New York: The American Historical Society, Inc., 1935. Vol. I.

Shinn, Charles Howard. *The Story of the Mine.* New York: D. Appleton and Company, 1896.

Spence, Clark C. *British Investments and the American Mining Frontier, 1860–1901.* Ithaca: Cornell University Press, 1958.

Trimble, William J. *The Mining Advance Into the Inland Empire.* Madison: University of Wisconsin, 1914.

Twain, Mark. *Roughing It.* New York: Rinehart and Company, 1953.

Wasson, Joseph. *Bodie and Esmeralda.* San Francisco: Spaulding, Barto & Co., 1878.

INDEX